ADVANCED
CALCULUS
WITH
LINEAR
ANALYSIS

ADVANCED CALCULUS
WITH
LINEAR ANALYSIS

Joseph R. Lee
Colorado School of Mines

Academic Press
New York and London

ACADEMIC PRESS, INC.
111 Fifth Avenue, New York, New York 10003

United Kingdom Edition published by
ACADEMIC PRESS, INC. (LONDON) LTD.
24/28 Oval Road, London NW1 7DD

LIBRARY OF CONGRESS CATALOG CARD NUMBER: 70-159611

AMS (MOS) 1970 Subject Classification 26-01

PRINTED IN THE UNITED STATES OF AMERICA

CONTENTS

PREFACE

The material in this book can best be described as advanced calculus from the point of view of linear spaces. It is the author's belief that just as operational techniques proved of great utility in the study of integral equations and related theories, so may they also be used to advantage in the study of traditional topics in advanced calculus. For example, at an early stage we discuss norm convergence in the space of continuous functions and show that the space is complete while still giving the traditional phraseology that the uniform limit of continuous functions is continuous.

Some of the advantages of this approach are the ease with which other convergence problems can be considered, the unifying aspect of linear spaces, and, as a kind of dividend to the student, some familiarity with the powerful tools of functional analysis. It should be emphasized, however, that this is not an attempt to shuffle together two decks of cards marked advanced calculus and functional analysis, but rather to use modern tools for the study of traditional topics.

The material covered in this book requires a thorough knowledge of calculus and some familiarity with topics in linear algebra and differential equations. Beyond that, the mathematical maturity of the students will determine the ease with which the material can be covered. No effort is made to have the presentation self-contained from the beginnings of set theory through the development of real numbers and functions. Results from elementary calculus are used as needed without specific reference. Thus, the student is assumed to be familiar with continuous functions when they are used in Chapter I as an example of linear spaces, but they are also studied in detail from a somewhat different point of view in Chapter IV. Concepts from topology are introduced as needed instead of in an introductory chapter. The Heine–Borel theorem, for example, appears in the study of continuous functions.

The choice of the few topics from calculus to be repeated has been dictated somewhat by experience of students' needs and somewhat by their suitability for inclusion in the study of linear spaces. For example, l'Hospital's rule is used freely without specific development in the text, but a concise review of infinite series is included—partly to give the student a better concept of the elements of the l_p spaces.

The completeness of various spaces is stressed until it is discovered that Riemann integrable functions do not form a complete linear space under any reasonable norm. Convergence properties for the Riemann integral are discussed in considerable detail without use of a norm, and this point may be the termination of a short course in advanced calculus. However, chapters on measure theory and Lebesgue integration are included so that a complete space of integrable functions is obtained.

ACKNOWLEDGMENTS

The author wishes to thank his students and colleagues for their aid in the preparation of this book. Special thanks go to Frederick M. Williams, Kempton L. Huehn, and Professors Ardel J. Boes and Donald C. B. Marsh. Also, the kind and expert work of the editorial and production staffs of Academic Press is acknowledged with grateful appreciation.

SUMMARY OF NOTATION
FOR LINEAR SPACES

Name of space	Elements	Norm
$C[a, b]$	$f(t)$ continuous for $a \leq t \leq b$	$\|f\| = \sup\limits_{a \leq t \leq b} \|f(t)\|$
$B[a, b]$	$f(t)$ bounded for $a \leq t \leq b$	$\|f\| = \sup\limits_{a \leq t \leq b} \|f(t)\|$
$C^{(1)}[a, b]$	$f(t)$ such that $f'(t)$ continuous for $a \leq t \leq b$	$\|f\| = \sup\limits_{a \leq t \leq b} \|f(t)\| + \sup\limits_{a \leq t \leq b} \|f'(t)\|$
E^n	$x = (\alpha_1, ..., \alpha_n)$	$\|x\| = \left(\sum\limits_{k=1}^{n} \|\alpha_k\|^2 \right)^{1/2}$
$l_p{}^n, 1 \leq p < \infty$	$x = (\alpha_1, ..., \alpha_n)$	$\|x\| = \left(\sum\limits_{k=1}^{n} \|\alpha_k\|^p \right)^{1/p}$
$l_\infty{}^n$	$x = (\alpha_1, ..., \alpha_n)$	$\|x\| = \sup\limits_{1 \leq k \leq n} \|\alpha_k\|$
$l_p, 1 \leq p < \infty$	$x = \{\alpha_k\}$ such that $\|x\|$ is finite	$\|x\| = \left(\sum\limits_{k=1}^{\infty} \|\alpha_k\|^p \right)^{1/p}$
l_∞	$x = \{\alpha_k\}$, bounded	$\|x\| = \sup\limits_{k} \|\alpha_k\|$

Name of space	Elements	Norm
(c)	$x = \{\alpha_k\}$, convergent	$\|x\| = \sup_k \|\alpha_k\|$
(c_0)	$x = \{\alpha_k\}$, convergent to zero	$\|x\| = \sup_k \|\alpha_k\|$
$\mathcal{L}_p$	Equivalence classes of measurable functions f such that f^p is Lebesgue integrable	$\|f\| = \int \|f\|^p$

LIST OF ELEMENTARY SYMBOLS[†]

Symbol	Definition
$x \in A$	x is an element of A
$A \cup B$	The set of all elements x such that $x \in A$ or $x \in B$ (or both)
$A \cap B$	The set of all elements x such that $x \in A$ and $x \in B$
$\displaystyle\bigcup_{k=1}^{n} A_k$	The set of all elements x such that x is an element of at least one A_k, $k = 1, 2, ..., n$
$\displaystyle\bigcup_{k=1}^{\infty} A_k$	The set of all elements x such that x is an element of at least one A_k, $k = 1, 2, ..., n, ...$
$\displaystyle\bigcap_{k=1}^{n} A_k$	The set of all elements x such that x is in every A_k, $k = 1, 2, ..., n$
$\displaystyle\bigcap_{k=1}^{\infty} A_k$	The set of all elements x such that x is in every A_k, $k = 1, 2, ..., n, ...$
$A \subset B$	Every element of A is an element of B
$A \supset B$	Every element of B is an element of A
$\displaystyle\sum_{k=1}^{n} f(k)$	$f(1) + f(2) + \cdots + f(n)$

† The symbols in this list have been used without definition in the text.

Symbol	*Definition*
$\displaystyle\sum_{k=1}^{\infty} f(k)$	$\displaystyle\lim_{n\to\infty} \sum_{k=1}^{n} f(k)$
$\displaystyle\prod_{k=1}^{n} f(k)$	$f(1)\cdot f(2)\cdot\ \cdots\ \cdot f(n)$
$\displaystyle\prod_{k=1}^{\infty} f(k)$	$\displaystyle\lim_{n\to\infty} \prod_{k=1}^{n} f(k)$
$\{x \mid P(x)\}$	The set of all x such that $P(x)$

1

FUNCTION
SPACES

1.1 linear spaces

The concept of a (single-valued) function, consisting of a domain,
a rule, and a range, is assumed to be familiar from elementary calculus.
Unless otherwise stated, we shall be concerned only with real-valued
functions of a real variable; that is, both the domain and range are
sets of real numbers.

If y is a function of x, our attention is often focused on the domain
and range spaces, and we investigate such questions as: When x
increases does y increase? What value of x makes y a maximum?
and If the time rate of change of y is proportional to y, when does
y double? Some of the most fruitful investigations in modern mathematics, however, have come about by concentrating on the rule, or
correspondence, rather than on the domain and range.

To this end we wish to consider a function f as a "point" in a
"space," a function space, and we will discuss such questions as:
How close is f to g? and Does f_n approach f as a limit? We will use
such notations as $f \in C[a, b]$ to mean that f is an element of the

space of functions continuous on $[a, b]$ as well as saying "let $f(x)$ be continuous for $x \in [a, b]$."

If the word "space" is well chosen, we can expect certain geometric concepts to carry over to our new spaces. In particular, we are concerned here with properties of linearity and distance; however, analogs of direction and angle also play an important role in more advanced studies.

1.1.1 *DEFINITION*   Let X be a set of elements (points) denoted by $f, g, h, \ldots$ or $x, y, z, \ldots$ and let $\alpha, \beta, \ldots$ be real numbers. Then X is a *real linear space* if the following axioms hold:

(1) $\alpha x + \beta y \in X$.
(2) $x + y = y + x$.
(3) $x + (y + z) = (x + y) + z$.
(4) There exists in X a unique element called the *zero element*, denoted by 0, such that $x + 0 = x$ for all $x \in X$.
(5) For each $x \in X$, there exists a unique element $(-x)$ in X such that $x + (-x) = (-x) + x = 0$.
(6) $\alpha(x + y) = \alpha x + \alpha y$.
(7) $(\alpha + \beta)x = \alpha x + \beta x$.
(8) $\alpha(\beta x) = (\alpha\beta)x$.
(9) $1x = x$.
(10) $0x = 0$.

It is not claimed that these axioms are independent; for example, the last part of (5) and all of (10) can be proved from the others. They include the familiar closure, commutative, associative, and distributive laws from vector algebra with which linear function spaces share a common heritage. It should also be noted that in Axioms (4), (5), and (10), the symbol "0" is used ambiguously to denote both the zero element of the linear space and the number 0. This should cause no trouble, however, because the context will make the meaning clear.

To interpret functions $f, g, h, \ldots$ as elements of a linear space it is necessary to define $f + g$, αf, and the zero element of the space. This is done by telling how the functions operate on a sample real number t. Thus, $(f + g)t = f(t) + g(t)$; $(\alpha f)t = \alpha f(t)$; and the zero element takes everything into zero, $0(t) = 0$. (Note the ambiguous use of

"0" again.) There are other operations, not necessarily defined for all linear spaces, which may be defined for real-valued functions of real variables. Thus, $(fg)t = f(t) \cdot g(t)$, provided t is in the domain of both f and g; $(1/f)t = 1/f(t)$ for t in the domain of f and $f(t) \neq 0$; $(f \circ g)t = f(g(t))$ for t in the domain of g and $g(t)$ in the domain of f.

Also, $\lim_{n \to \infty} f_n = f$ means $\lim_{n \to \infty} f_n(t) = f(t)$ for t in the domains of f_n $(n = 1, 2, \ldots)$ and f. Note that $\{f_n(t)\}$ is a sequence of numbers in the union of the ranges of f_n. Other types of convergence will be discussed later, and, to avoid confusion, this will be referred to as *pointwise convergence*.

We shall have occasion to refer to many function spaces repeatedly, so it is convenient to have the following notations available. (See Summary of Notation for Linear Spaces, p. *xi*.)

1.1.2 *NOTATION* The space of functions continuous for $a \leq t \leq b$ is denoted by $C[a, b]$.

1.1.3 *NOTATION* The space of functions bounded for $a \leq t \leq b$ is denoted by $B[a, b]$.

1.1.4 *NOTATION* The space of functions with bounded continuous first derivatives for $a \leq t \leq b$ is denoted by $C^{(1)}[a, b]$.

It should be noted that the domain in all three cases is a closed, bounded interval. We may have occasion to use *examples* where the domain is open, or half-open and half-closed, and perhaps unbounded. The usual notations will apply. Thus, for example, $B[a, \infty)$ means the space of all functions bounded for $a \leq t < \infty$. In the statements of *theorems*, however, care should be taken to note that the interval is usually bounded and closed. Thus it is true that $C[a, b] \subset B[a, b]$, but not necessarily true that $C(a, b) \subset B(a, b)$.[†]

PROBLEMS

1. Verify that $C[a, b]$, $B[a, b]$, and $C^{(1)}[a, b]$ are linear spaces.

[†] In Section 4.3 the terms "closed" and "bounded" will be defined for sets of points other than intervals. The definitions given here can then be extended to cases like $C[T]$ to mean the space of functions that are continuous for $t \in T$, a closed, bounded set of numbers.

2.   Determine which of the following are linear spaces. If they are not, which axioms are violated?

(a)   All functions defined for $0 \leq t \leq 4$ whose derivatives exist and are equal to 0 for $t = 1, 2, 3$.

(b)   All functions defined for $0 \leq t \leq 4$ whose derivatives exist and are equal to 1 for $t = 1, 2, 3$.

(c)   All functions with no zero for $0 \leq t \leq 4$.

3.   If $f(t) = \sin^{-1} t$, $g(t) = \cos^{-1} t$, where $0 \leq t \leq 1, 0 \leq f(t) \leq \pi/2$, and $0 \leq g(t) \leq \pi/2$, show that $(f + g)t = \pi/2$ for all t. Thus $f + g \in C^{(1)}[0, 1]$. (Why?)

4.   In Problem 3, is $f \in C^{(1)}[0, 1]$? Is $g \in C^{(1)}[0, 1]$?

5.   Prove the following: if $f \in C^{(1)}[a, b]$, and $f + g \in C^{(1)}[a, b]$, then $g \in C^{(1)}[a, b]$.

In Problems 6–8, $f(t) = t^4 - t^2$, $g(t) = 1 - t^2$, and $h(t) = \sqrt{t}$. For each problem describe the required function by telling how it operates on t, and give the domain and range.

6. $g \circ h$.   **7.** $h \circ (f + g)$.   **8.** f/g.

9.   If $f(t) = \sin 2t$, $g(t) = \cos 2t$, find $f \circ g$, $f + g$, and f/g. Give their domains and ranges.

10.   Let $f(t) = \dfrac{1}{\sqrt{4 - t^2}}$ and $g(t) = 2 \cos t$. If $F = f \circ g$, show

$F(t) = \frac{1}{2} | \csc t |$.

11.   If $F(t) = \cos t$ and $g(t) = \sec t$, find a function f such that $F = f \circ g$.

In problems 12–15, $\lim_{n \to \infty} f_n(t) = f(t)$. Find $f(t)$.

12. $f_n(t) = t^n$, $0 \leq t \leq 1$.

13. $f_n(t) = \sqrt[n]{\sin t}$, $0 \leq t \leq \pi/2$.

14. $f_n(t) = \dfrac{nt}{1 + n^2 t^2}$, $0 \leq t < \infty$.

15. $f_n(t) = \dfrac{1}{n^t}$, $1 \leq t < \infty$.

In Problems 16–20, find the largest interval so that the given function is an element of $C^{(1)}$ of that interval. (*Note:* the end points may be $+\infty$ or $-\infty$.)

16. $f(t) = (t - 1)^{1/3} + (4 - t)^{3/2}$.

17. $f(t) = \ln \sqrt{1 - t^2}$.

18. $f(t) = \ln(t^2 + 4)$.

19. $f(t) = t^2 \sqrt{3 - 4t}$.

20. $f(t) = \sqrt{8t - \dfrac{1}{t^2}}$.

Among the three spaces defined thus far, we have $C^{(1)}[a, b] \subset C[a, b] \subset B[a, b]$. (The student is probably familiar with these results from elementary calculus; for proofs, see Theorems 4.1.6 and 5.1.1.) In Problems 21–30, find the smallest space (or possibly none), for the interval given, of which the given function is an element.

21. $f(t) = |t|$, $-1 \le t \le 1$.

22. $f(t) = \sqrt{1 - t^2}$, $-1 \le t \le 1$.

23. $f(t) = \dfrac{1}{\sqrt{1 - t^2}}$, $-1 < t < 1$.

24. $f(t) = \sin t + \cos t$, $-\pi \le t \le \pi$.

25. $f(t) = \tan t^2$, $-\pi \le t \le \pi$.

26. $f(t) = t \sin \dfrac{1}{t}$, $-\dfrac{1}{\pi} \le t \le \dfrac{1}{\pi}$.

27. $f(t) = t \sqrt{1 - t^2}$, $-1 \le t < 1$.

28. $f(t) = \begin{cases} \dfrac{e^t - e^{-t}}{\sin t}, & -\dfrac{\pi}{2} \le t < 0 \quad \text{and} \quad 0 < t \le \dfrac{\pi}{2}, \\ 2, & t = 0. \end{cases}$

(*Hint:* use l'Hospital's rule to determine if $f \in C[-\pi/2, \pi/2]$.

What about $f \in C^{(1)}[-\pi/2, \pi/2]$?)

29. $f(t) = \begin{cases} \sec t - \tan t, & 0 \leq t < \dfrac{\pi}{2} \text{ and } \dfrac{\pi}{2} < t \leq \pi, \\ \\ 0, & t = \dfrac{\pi}{2} \end{cases}$

30. $f(t) = \begin{cases} t \ln (\sin t), & 0 < t \leq \dfrac{\pi}{2}, \\ \\ 1, & t = 0. \end{cases}$

1.2 normed spaces

Another useful concept in linear analysis is the generalization of distance, starting with the distance of a point from the origin. The familiar definition of $|t| = +t$ if $t \geq 0$, $|t| = -t$ if $t < 0$ means intuitively that $|t|$ is the distance of the point t on the real line from the origin without regard to direction. We have already seen this concept extended to complex numbers; for, if $t = u + iv$ ($i^2 = -1$) is a complex number, $|t| = \sqrt{u^2 + v^2}$, which is also the distance of the point (u, v) from the origin. It is easily seen that in both the real and complex cases, $|t_2 - t_1|$ is the distance between the points t_1 and t_2 without regard to direction.

We wish to generalize this idea of distance for points in any linear space:

1.2.1 *DEFINITION* The *norm* of $x \in X$, written $\| x \|$, is a real number associated with the element x satisfying the following:

(1) $\| 0 \| = 0$, $\| x \| > 0$ if $x \neq 0$.
(2) $\| x + y \| \leq \| x \| + \| y \|$.
(3) $\| \alpha x \| = | \alpha | \| x \|$.

Thus the norm of an element x can be considered the "distance" of the element x from the "origin" 0, and $\| x - y \|$ is the "distance between x and y."

The question naturally arises as to whether there could be other norms satisfying the above conditions for the same set of points. As we shall see, the answer is yes. Thus a given linear space may have different norms imposed on it (see Problem 3). When this is done, it may be necessary to distinguish between different norms in the same set of points by $\| x \|_1$ and $\| x \|_2$, for example. Unless otherwise stated, however, *the* norm in a given space will be the one stated in the definitions below.

Before proceeding to definitions of specific norms, we recall the following definitions from calculus:

1.2.2   DEFINITION M is an *upper bound* of a set S of real numbers if $t \leq M$ for every $t \in S$.

1.2.3   DEFINITION   The *supremum*, or *least upper bound*, of a set S of real numbers, written sup S or lub S, is a number u with the following properties:

(1) For every $t \in S$, $t \leq u$, that is, u is an upper bound of S.
(2) for every $\epsilon > 0$, there is a $t \in S$ such that $t > u - \epsilon$.

The lower bound and infimum are defined in a similar manner. (See Problem 9.)

It should be recalled that the supremum may or may not belong to the set S. Thus $S = \{t \mid t^2 \leq 4\}$ has sup $S = 2$, and $2 \in S$. If $R = \{t \mid t^2 < 4\}$, sup $R = 2$, and $2 \notin R$. If $Q = \{t \mid t^2 \leq 2$, and t is rational$\}$, sup $Q = \sqrt{2}$, and $\sqrt{2} \notin Q$. If sup S is an element of S, it may be written max S Thus in the examples above, max $S = 2$, but there is no maximum for R or Q.

We shall now return to the definitions of norms in the function spaces already discussed:

1.2.4   DEFINITION   If $f \in C[a, b]$,

$$\| f \| = \sup_{a \leq t \leq b} | f(t) |.$$

1.2.5   DEFINITION   If $f \in B[a, b]$,

$$\| f \| = \sup_{a \leq t \leq b} | f(t) |.$$

1.2.6 DEFINITION   If $f \in C^{(1)} [a, b]$,

$$\| f \| = \sup_{a \le t \le b} |f(t)| + \sup_{a \le t \le b} |f'(t)|.$$

Since we shall be discussing many normed linear spaces, the symbol $\| f \|$ will have different meanings depending on what space f is in. In any given problem the norm remains the same, however, so no ambiguity arises.

When we are dealing with different kinds of convergence in the same discussion, we shall identify the norm to be considered by such statements as "$f_n \to f$ in the $C^{(1)}$ norm," or "$f_n \to f$ in the sup norm," or "$\| f_n \| \to 0$, where $\| \cdot \|$ is the norm in $C^{(1)}$."

PROBLEMS

1. If t is (a) a real number, or (b) a complex number, show that $|t|$ is a norm. In each case, where do points lie for which $|t| \le 1$?

2. If x and y are (a) real numbers, or (b) complex numbers, where do x and y lie if equality holds in Definition 1.2.1, part (2)?

3. If $t = u + iv$ is a complex number, define $\| t \|_1 = |u| + |v|$. Is this a norm? Where do points lie for which $\| t \|_1 = 1$?

4. Let
$$f(t) = \begin{cases} 4t - t^3 & \text{if } -2 \le t < 0, \quad 0 < t \le 2, \\ 0 & \text{if } t = 0. \end{cases}$$
Let S be the range of values of $f(t)$ for $t \in [-2, 2]$. Find sup S. Is this an element of S?

5. Let $f(t) = t^3 - 3t^2 + 2t$, and let R be the range of values of $f(t)$ for $t \in [0, 4]$ and t rational. Find sup R. Is this an element of R?

6. Prove that if sup S is finite, then it is unique.

7. If $f \in C[a, b]$ or $f \in B[a, b]$, show that $\sup_{a \le t \le b} |f(t)|$ is a norm.

8. If $f \in C^{(1)}[a, b]$, why is just $\sup_{a \leq t \leq b} |f'(t)|$ not a norm?

9. Give definitions for the lower bound and the infimum, or greatest lower bound, of a set of numbers similar to Definitions 1.2.2 and 1.2.3.

10. In Section 1.1, Problem 12, the function f was found as the pointwise limit of $\{f_n\}$, that is,

$$\lim_{n \to \infty} [f_n(t) - f(t)] = 0 \qquad \text{for} \quad 0 \leq t \leq 1.$$

Does $\lim_{n \to \infty} \|f_n - f\| = 0$, where $\| \cdot \|$ is the norm in $C[0, 1]$?

11. In Section 1.1, Problem 13, does $\lim_{n \to \infty} \|f_n - f\| = 0$, where $\| \cdot \|$ is the norm in $C[0, \pi/2]$?

12. In Section 1.1, find $\|f\|$ in Problems 21–30.

13. A *metric space* is a collection of points, x, y, z, $\ldots$, with a "distance function," or *metric* $d(x, y)$ defined on pairs of points with the following properties:

(a) $d(x, y) = d(y, x)$,
(b) $d(x, y) \leq d(x, z) + d(y, z)$,
(c) $d(x, x) = 0$,
(d) If $d(x, y) = 0$, then $x = y$.

If we define $d(x, y) = \|x - y\|$, show that every normed linear space is a metric space.

14. In a metric space show that $d(x, y) \geq 0$.

15. Let $x = (x_1, x_2)$ and $y = (y_1, y_2)$ be two points in two-dimensional Euclidean space. Show that the distance formula from analytic geometry is a metric.

16. Discuss the validity of the following statements: "every subset of points of a metric space is a metric space"; and "every subset of points of a linear normed space is a linear normed space."

17. Let X be a collection of points x, y, z, $\ldots$, and let $d(x, y) = 1$ if $x \neq y$, $d(x, y) = 0$ if $x = y$. Is X a metric space?

1.3 some inequalities

Before proceeding to infinite sequences and series we shall consider some examples of normed linear spaces whose elements are finite sequences, or n-tuples. Since we have been following the analogy between geometric spaces and general linear spaces, a word about notation is in order. Instead of the familiar $P(x, y)$ to denote a point with abscissa x and ordinate y, we wish to use x as a point in our space. So, for example, $x:(\alpha_1, \alpha_2)$, or $x = (\alpha_1, \alpha_2)$, will be a point in two-dimensional space with coordinates α_1 and α_2. The advantage of this notation in its extension to higher dimensions is obvious.

We shall define several spaces whose elements (points) are n-tuples and, later, infinite sequences. If these are to satisfy the axioms for linear spaces, the element 0 and the operations + and scalar product must be defined. In two-dimensional Euclidean space the zero element is the origin, $0 = (0, 0)$, and addition of two points is performed by adding the corresponding coordinates, $(\alpha_1, \alpha_2) + (\beta_1, \beta_2) = (\alpha_1 + \beta_1, \alpha_2 + \beta_2)$, and $\alpha(\alpha_1, \alpha_2) = (\alpha\alpha_1, \alpha\alpha_2)$. Similar definitions hold for n-tuples and infinite sequences and will not be repeated for each space.

1.3.1 *NOTATION*   The space E^n is the space of ordered n-tuples $x = (\alpha_1, \ldots, \alpha_n)$, where α_i $(i = 1, \ldots, n)$ is a real number. The norm is

$$\| x \| = \left(\sum_{k=1}^{n} | \alpha_k |^2 \right)^{1/2} = (| \alpha_1 |^2 + \cdots + | \alpha_n |^2)^{1/2}.$$

We call E^n n-dimensional Euclidean space.

1.3.2 *NOTATION*   The space $l_p{}^n$ is defined for positive integers n and real numbers p with $1 \leq p < \infty$. The elements are n-tuples $x = (\alpha_1, \ldots, \alpha_n)$ of real numbers with the norm

$$\| x \| = \left(\sum_{i=1}^{n} | \alpha_i |^p \right)^{1/p}.$$

1.3.3 *NOTATION* The space $l_\infty{}^n$ is the space of all ordered n-tuples $x = (\alpha_1, \ldots \alpha_n)$ of real numbers with the norm

$$\| x \| = \sup_{1 \leq i \leq n} | \alpha_i |.$$

Note that $l_2{}^n$ is the same as E^n and that $l_1{}^2$ is the same as the space in Problem 3, Section 1.2. Note also that the three spaces defined above all have the same points but different norms. They are *pointwise* equivalent but with different norms.

Since the norm plays such an important part in describing a linear space, the student might well ask why the particular norms were chosen in the spaces that we have discussed so far. Three requirements that must be kept in mind in defining a norm are obviously the properties described in Definition 1.2.1. However, the elements in $C^{(1)}[a, b]$ obviously form a subset of the elements in $C[a, b]$, so if these were the only requirements, the same norm could have been used in both cases. We shall return to this important question in Chapter III, because it introduces one of the important features that make linear spaces useful in the study of classical function theory.

In order to prove that the three spaces defined above are indeed normed linear spaces (see Problem 1), some special inequalities are needed:

1.3.4 *LEMMA* For $x > 1,\ 0 < m < 1$, we have

$$x^m - 1 < m(x - 1).$$

Proof For $x > 0, F(x) = x^m - 1 - m(x - 1)$ is continuous with a maximum of 0 at $x = 1$. Therefore, for $x > 1, F(x) < 0$ and the inequality follows. ▌

In the following two theorems, the symbol $\sum$ stands for the finite sum from $k = 1$ to $k = n$. The student should note in the proofs, however, that it could also be interpreted as an infinite sum (provided the series converges) or as an integral [where a_k is replaced by $f(t)$, for example].

1.3.5   THEOREM   (Hölder's inequality) [†]

$$| \sum a_k b_k | \leq (\sum | a_k |^p)^{1/p} (\sum | b_k |^q)^{1/q} \quad \text{where } \frac{1}{p} + \frac{1}{q} = 1.$$

Proof   In Lemma 1.3.4 let $x = a/b$ with $a > b > 0$, and multiply both sides of the equation by b. Then,

$$a^m b^{1-m} < b + m(a - b).$$

Now, let $m = \alpha$, $1 - m = \beta$, so that $\alpha + \beta = 1$. Then, $a^\alpha b^\beta < a\alpha + b\beta$. Since this is symmetric in a, α and b, β, it also holds if $b > a$; there is equality if $a = b$. Now, let A_k and B_k be positive. Then

$$\sum \left[\left(\frac{A_k}{\sum A_k} \right)^\alpha \left(\frac{B_k}{\sum B_k} \right)^\beta \right] \leq \sum \left(\alpha \frac{A_k}{\sum A_k} + \beta \frac{B_k}{\sum B_k} \right)$$
$$= \alpha + \beta = 1.$$

Thus

$$\sum A_k^\alpha B_k^\beta \leq (\sum A_k)^\alpha (\sum B_k)^\beta.$$

Now, let $\alpha = 1/p$, $A_k = | a_k |^p$, $\beta = 1/q$, $B_k = | b_k |^q$, and the inequality follows.   ∎

In case $p = q = 2$, Hölder's inequality is known as Schwarz's inequality. (See Problem 2.)

1.3.6   THEOREM   (Minkowski's inequality) Let $p > 1$, $(1/p) + (1/q) = 1$. Then,

$$(\sum | a_k + b_k |^p)^{1/p} \leq (\sum | a_k |^p)^{1/p} + (\sum | b_k |^p)^{1/p}.$$

Proof   Note   that $| a_k + b_k |^p = | a_k + b_k | \cdot | a_k + b_k |^{p-1}$. Then,

$$\sum | a_k + b_k |^p \leq \sum | a_k | \cdot | a_k + b_k |^{p-1}$$
$$+ \sum | b_k | \cdot | a_k + b_k |^{p-1}$$
$$\leq (\sum | a_k |^p)^{1/p} (\sum | a_k + b_k |^p)^{1/q}$$
$$+ (\sum | b_k |^p)^{1/p} (\sum | a_k + b_k |^p)^{1/q}$$

[†] Many proofs of this and the following inequality are known, some suggested in the problems.

by Hölder's inequality. Now divide both sides of this in-equality by $(\sum |a_k + b_k|^p)^{1/q}$, and the inequality in the theorem follows. ∎

PROBLEMS

1. Using Minkowski's inequality show that the norm defined for $l_p{}^n$ satisfies Definition 1.2.1.

2. Show that

$$\left(\sum_{k=1}^{n} a_k{}^2\right)\left(\sum_{k=1}^{n} b_k{}^2\right) - \left(\sum_{k=1}^{n} a_k b_k\right)^2 = \sum_{1 \le k < j \le n} (a_k b_j - a_j b_k)^2 \ge 0.$$

and thus establish Schwarz's inequality.

3. Using Problem 2, establish Minkowski's inequality for $p = q = 2$. (*Hint*: state Minkowski's inequality for this special case, square both sides, simplify, and reverse steps.)

4. Prove Schwarz's inequality as follows: define $\sum_{k=1}^{n} (a_k - \lambda b_k)^2$ as a function of λ that cannot change sign. Obtain this in the form $A\lambda^2 + 2B\lambda + C$. Then, $B^2 - AC < 0$. (Why?) This gives the desired inequality.

5. Interpret $l_1{}^3$ as Euclidean 3-space, but with the norm of Notation 1.3.2. If $x = (\alpha_1, \alpha_2, \alpha_3)$, describe the set of points for which $\| x \| = 1$.

6. Let $x_i = \{\alpha_j{}^{(i)}\}$, where

$$\alpha_j{}^{(i)} = \frac{1}{j} + \frac{1}{j^i}, \qquad i = 1, \ldots, n, \quad j = 1, \ldots, k.$$

Fill in the ?'s in the following table.

$$x_1 = (?, ?, ?, \ldots, ?),$$
$$x_2 = (?, ?, ?, \ldots, ?),$$
$$x_3 = (?, ?, ?, \ldots, ?),$$
$$\vdots$$
$$x_n = (?, ?, ?, \ldots, ?).$$

(Note that this is a finite array.)

7. In Problem 6 consider $x_i \in l_1^k$, $i = 1, \ldots, n$. Find $\| x_1 \|$, $\| x_2 \|$, and $\| x_n \|$.

8. In Problem 6 with $x_i \in l_1^k$, find $\| x_1 - x_2 \|$ and $\| x_m - x_n \|$.

9. Do Problem 7 with $x_i \in l_2^k$.

10. Do Problem 8 with $x_i \in l_2^k$.

11. Do Problem 7 with $x_i \in l_\infty^k$.

12. Do Problem 8 with $x_i \in l_\infty^k$.

II

SEQUENCE SPACES AND INFINITE SERIES

2.1 sequences

The n-tuples considered in Section 1.3 can be extended to infinite sequences, or, alternately, we could consider n-tuples as infinite sequences with all but a finite number of the components zero. The student is already familiar with sequences and series of real numbers from elementary calculus and, to some extent, with sequences of functions. We shall enlarge upon these concepts and consider their extension to sequences and series in general normed linear spaces. Some of these spaces have sequences as their elements, but the idea of sequences of sequences is a novelty that causes no real trouble.

In the following, then, the words "number," "point," and "element" are to be considered interchangeable unless the context implies some limitation. (For example, $x_1 < x_2$ has meaning for real numbers, but not for complex numbers or for functions.) If a proposition involves the absolute value of a number, this can be interpreted as the norm of an element. Thus, if x is a real number, we may encounter $|x|$; and if x in the same proposition is to be interpreted as an element of a general linear space, it must be understood that $|x|$ becomes $\|x\|$.

2.1.1 *DEFINITION* An *infinite sequence* is an ordered set of elements that can be put into a one-to-one correspondence with the positive integers.

If a sequence is to be of much interest, there should be some means of finding the general, or nth, term. When this is true, the three elements (domain, rule, and range) of a function are present so that a sequence may be considered a function from the positive integers to a set of elements. For example, the formula

$$a_n = \frac{n}{2^n - 1}$$

gives the sequence 1, 2/3, 3/7, 4/15, ... of real numbers, while a sequence of functions $\{f_n\}$ may be described by telling how the nth function operates on a number t, such as

$$f_n(t) = \frac{t^n}{1 + t^n}.$$

2.1.2 *DEFINITION* The sequence $\{a_n\}$ is said to *converge* to a finite number a if $\lim_{n\to\infty} a_n = a$; that is, if given $\epsilon > 0$, there exists an N such that $|a_n - a| < \epsilon$ if $n > N$. If no such a exists, the sequence diverges. If for every $M > 0$ there is an N such that $|a_n| > M$ when $n > N$, we write $\lim_{n\to\infty} a_n = \infty$. If it is necessary to distinguish between $+\infty$ and $-\infty$, we require $a_n > M$ in the first case, and $a_n < -M$ in the second, when $n > N$.

A little reflection indicates that there are in general two ways in which a sequence of real numbers may diverge; we may have $\lim_{n\to\infty} a_n = \infty$, or the numbers in $\{a_n\}$ may oscillate in some way to prevent $|a_n - a|$ remaining small for any a. The concept of a monotone sequence will make these ideas more precise.

2.1.3 *DEFINITION* A sequence $\{a_n\}$ of real numbers is *monotone increasing* if $a_n < a_{n+1}$ for all n; it is *montone nondecreasing* if $a_n \le a_{n+1}$; it is *monotone decreasing* if $a_n > a_{n+1}$; it is *monotone nonincreasing* if $a_n \ge a_{n+1}$. A sequence satisfying any one of the above conditions is said to be a *monotone* sequence.

We state a theorem here for reference but postpone the proof until Chapter III. The student should determine at that time that no circular reasoning has been involved.

2.1.4 *THEOREM* Every bounded monotone sequence of real numbers converges.

Proof See Theorem 3.1.3. ∎

If $a_n = 1 - 2^{-n}$, the sequence $\{a_n\}$ is monotone increasing and is bounded above by 1; so $\{a_n\}$ will converge. Obviously, $\lim_{n \to \infty} a_n = 1$. On the other hand, if $b_n = 1 + (-1)^n 2^{-n}$ and $c_n = (-1)^n(1 - 2^{-n})$, $\{b_n\}$ and $\{c_n\}$ are not monotone sequences; $\lim_{n \to \infty} b_n = 1$, but $\lim_{n \to \infty} c_n$ does not exist.

The last example suggests the possibility of describing other kinds of limits:

2.1.5 *DEFINITION* Let $\{a_n\}$ be a sequence of real numbers. Let a be a number with the property that for every $\epsilon > 0$ an infinite number of a_n's are greater than $a - \epsilon$ while only a finite number exceed $a + \epsilon$. Then a is called the *limit superior* of $\{a_n\}$, and we write $a = \lim \sup_{n \to \infty} a_n$.

A similar definition holds for $\lim \inf_{n \to \infty} a_n$, or limit inferior.

If we allow ∞ as an acceptable limit, $\lim \sup$ always exists (see Problem 10, Section 3.1). In the example above, $\lim \sup_{n \to \infty} c_n = +1$, and $\lim \inf_{n \to \infty} c_n = -1$.

The example

$$a_n = \frac{1}{n} + \sin \frac{n\pi}{2}$$

suggests still another kind of limit for the sequence $\{a_n\}$. Obviously an infinite number of points in the sequence cluster about 0, but that is neither the limit, the limit superior, nor the limit inferior. The following definition provides a name for this kind of point:

2.1.6 DEFINITION The point a is a *limit point*, or *cluster point*, of the sequence $\{a_n\}$ if every interval $(a - \epsilon, a + \epsilon)$ contains points of $\{a_n\}$ other than a itself.

(We leave for an open argument the question of whether the sequence 0, 1, 0, 1, 0, 1, . . . has any limit points.)

PROBLEMS

1. Give a definition of limit inferior of a sequence similar to that in Definition 2.1.5 for limit superior.

In Problems 2–13, a_n is given. For each problem write out several terms of the sequence and find $\lim_{n \to \infty} a_n$ if it exists; if not, find $\lim \sup_{n \to \infty} a_n$, $\lim \inf_{n \to \infty} a_n$, and all limit points.

2. $a_n = \dfrac{5n}{2^n + 1}$, $n = 1, 2, 3, \ldots.$

3. $a_n = \dfrac{2^n}{3^n + 1}$, $n = 1, 2, 3, \ldots.$

4. $a_n = \ln\left[1 + (-1)^n \dfrac{1}{n}\right]$, $n = 2, 3, 4, \ldots.$

5. $a_n = \ln\left[2^{(-1)^n} + \dfrac{1}{n}\right]$, $n = 1, 2, 3, \ldots.$

6. $a_n = \dfrac{\ln(n + 1) - \ln(n - 1)}{n^{-1}}$, $n = 2, 3, 4, \ldots.$

7. $a_n = (-1)^n\left(1 + \dfrac{1}{n}\right) + \sin\dfrac{n\pi}{2}$, $n = 1, 2, 3, \ldots.$

8. $a_n = (-1)^n \left(1 - \dfrac{1}{n}\right) + \cos\dfrac{n\pi}{2}, \qquad n = 1, 2, 3, \ldots .$

9. $a_n = (-1)^n \left(1 - \dfrac{1}{n}\right) + (-1)^{n+1}\left(1 + \dfrac{1}{n}\right), \qquad n = 1, 2, 3, \ldots .$

10. $a_n = 3 \sin\dfrac{n\pi}{2} + 2 \cos\dfrac{n\pi}{2}, \qquad n = 1, 2, 3, \ldots .$

11. $a_n = 2^{(-1)^n} + (-1)^n \left(2 + \dfrac{1}{n}\right), \qquad n = 1, 2, 3, \ldots .$

12. $a_n = 2^{(-1)^n} + 3^{(-1)^{n+1}}\left(1 + \dfrac{1}{n}\right), \qquad n = 1, 2, 3, \ldots .$

13. $a_n = \dfrac{1}{n}\left(\sin\dfrac{n\pi}{2} + \cos\dfrac{n\pi}{2}\right), \qquad n = 1, 2, 3, \ldots .$

14. Prove that if $\lim \sup_{n\to\infty} a_n = \lim \inf_{n\to\infty} a_n$, then $\lim_{n\to\infty} a_n$ exists and equals the common value of the other two.

15. Give an example in which the limit superior of a sequence is $-\infty$.

16. Show that the limit of a sequence of positive terms cannot be negative.

17. For the sequence $1, 3, 5, \ldots$, find two different formulas for a_n and thus two different possibilities for a_4.

18. If $a_n = n^2$, show that $a_{n+1} - a_n = (n + 1) + n$, and interpret the result geometrically.

2.2 sequence spaces

In many ways a convergent sequence behaves like a finite n-tuple. We therefore have spaces similar to those described in Section 1.3.

2.2.1   *NOTATION*   The space l_p is defined for $1 \leq p < \infty$ as the linear space of sequences $x = \{\alpha_k\}$ of real numbers for which $(\sum_{k=1}^{\infty} |\alpha_k|^p)^{1/p}$ is finite. The norm is

$$\| x \| = \left(\sum_{k=1}^{\infty} |\alpha_k|^p \right)^{1/p}.$$

2.2.2   *NOTATION*   The space l_∞ is the linear space of all bounded sequences $x = \{\alpha_k\}$ of real numbers. The norm is

$$\| x \| = \sup_k |\alpha_k|.$$

2.2.3   *NOTATION*   The space (c) is the linear space of all convergent sequences $x = \{\alpha_k\}$ of real numbers. The norm is

$$\| x \| = \sup_k |\alpha_k|.$$

2.2.4   *NOTATION*   The space (c_o) is the linear space of all sequences $x = \{\alpha_k\}$ of real numbers converging to zero. The norm is

$$\| x \| = \sup_k |\alpha_k|.$$

Addition and scalar multiplication of elements in these spaces are performed as in the spaces of n-tuples, and the zero element is $0 = (0, 0, 0, \ldots)$. It should be noted that similar sequence spaces may be constructed with complex numbers as components. In this case, l_2 is a representation of Hilbert space.

PROBLEMS

1. Of which spaces $[l_\infty, (c), \text{ and } (c_0)]$ are the sequences in Problems 2–13 in Section 2.1 elements?

2. Prove that $(c) \subset l_\infty$.

3. Prove that l_∞, (c), and (c_0) are linear spaces.

4. Let (c_1) be the set of sequences convergent to 1. Why is (c_1) not a linear space?

5. Prove that the sup norm used in l_∞, (c), and (c_0) is actually a norm.

6. Outline the proof that l_p is a normed linear space.

7. Let $x_1 = \{\alpha_k{}^{(1)}\}$, $x_2 = \{\alpha_k{}^{(2)}\}$, $\ldots$, $x_n = \{\alpha_k{}^{(n)}\}$ for $k = 1$, 2, $\ldots$, where

$$\alpha_k{}^{(n)} = \frac{k^{n-1} + 1}{k^n} = \frac{1}{k} + \frac{1}{k^n}.$$

Find $x = \lim_{n \to \infty} x_n$. Show $x_n \in (c_0)$ and $x \in (c_0)$.

8. Let $x_n = \{\alpha_k{}^{(n)}\}$, where

$$\alpha_k{}^{(n)} = \frac{nk + k + n}{nk}$$

for $n = 1$, 2, $\ldots$, $k = 1$, 2, $\ldots$. Find $x = \lim_{n \to \infty} x_n$. Show $x_n \in (c)$ and $x \in (c)$ but $x_n \notin (c_0)$ and $x \notin (c_0)$.

9. The elements $\{e_i\}$ in the space (c_0) form a *basis* if for every $x \in (c_0)$ there is a unique set of real numbers $\{a_i\}$ such that $\lim_{n \to \infty} \| x - \sum_{i=1}^n a_i e_i \| = 0$. Find a basis for (c_0). Will this also serve as a basis for (c)?

Prove or disprove (by giving a counterexample) the statements in Problems 10–15.

10. If $\lim_{n \to \infty} | a_n | = 1$, $\lim_{n \to \infty} a_n = 1$.

11. If $\{a_n\} \in (c_0)$, $\{| a_n |\} \in (c_0)$.

12. If $\lim_{n \to \infty} | a_n | = 1$, then $\lim \sup_{n \to \infty} a_n = 1$, or $\lim \inf_{n \to \infty} a_n = -1$, or both.

13. If $\lim_{n \to \infty} a_n = 1$, $\lim_{n \to \infty} | a_n | = 1$.

14. If $\{\alpha_n\} \in l_\infty$, then $\left\{ \frac{1}{n} \alpha_n \right\} \in (c_0)$.

15. If $\{\alpha_n\} \in (c_0)$, then $\{n \, \alpha_n\} \in l_\infty$.

2.3 infinite series—tests for convergence

The norm in l_p shows that for many infinite sequences we wish to discuss the associated infinite series. It is assumed that the student is familiar with elementary definitions and properties of infinite series of constant terms, but some of these are repeated here for convenience.

2.3.1 *DEFINITION*   An *infinite* *series* is an indicated sum

$$a_1 + a_2 + \cdots + a_n + \cdots = \lim_{n \to \infty} \sum_{k=1}^{n} a_k = \sum_{k=1}^{\infty} a_k.$$

With every series there are associated two sequences. One is simply the sequence $\{a_k\}$ whose components are terms of the series. Thus an element $x = \{\alpha_k\} \in l_p$ gives rise to a series which, in fact, is used to define $\| x \|$. The other sequence that concerns us is the sequence $\{A_n\}$ of partial sums, where $A_n = \sum_{k=1}^{n} a_k$.

2.3.2 *DEFINITION*   The series $\sum_{k=1}^{\infty} a_k$ *converges* to A if the sequence $\{A_n\}$ of partial sums converges to A. If no such A exists, the series diverges.

2.3.3 *DEFINITION*   The   series $\sum_{k=1}^{\infty} a_k$ *converges*   *absolutely* if $\sum_{k=1}^{\infty} | a_k |$ converges.

These definitions are obvious ones and should settle the matter of convergence or divergence of infinite series very easily. Unfortunately, obtaining the partial sums A_n in closed form is usually not an easy matter, so other forms of attack must be devised. A few cases in which Definition 2.3.2 *is* applied directly are shown in the following theorems and examples:

2.3.4 *THEOREM*   Let $a + ar + ar_2 + \cdots + ar^{n-1} + \cdots$ $(a \neq 0)$ be a geometric series with common ratio r. Then, if $r \neq 1$,

$$A_n = \frac{a - ar^n}{1 - r}.$$

Proof

$$A_n = a + ar + \cdots + ar^{n-1},$$

$$rA_n = ar + ar^2 + \cdots + ar^n.$$

The result follows by subtracting and dividing by $(1 - r)$. ∎

2.3.5 *COROLLARY* If $0 \leq r < 1$, the geometric series converges and $A = a/(1 - r)$. If $r \geq 1$, the geometric series diverges.

Proof For $r < 1$, and $r > 1$, write

$$A_n = \frac{a}{1 - r} - \frac{ar^n}{1 - r}$$

and the result follows. For $r = 1$, $A_n = na$ which approaches ∞ as $n \to \infty$. ∎

2.3.6 *EXAMPLE* The series

$$\sum_{k=1}^{\infty} \frac{1}{k(k + 1)}$$

converges because

$$A_n = \sum_{k=1}^{n} \frac{1}{k(k + 1)} = \sum_{k=1}^{n} \left(\frac{1}{k} - \frac{1}{k + 1} \right)$$

$$= 1 - \frac{1}{n + 1} \to 1 \qquad \text{as} \quad n \to \infty.$$

2.3.7 *THEOREM* If $\sum_{k=1}^{\infty} a_k$ converges, $\lim_{n \to \infty} a_n = 0$.

Proof Since $a_n = A_n - A_{n-1}$, then $\lim_{n \to \infty} a_n = A - A = 0$. ∎

This theorem is often applied in its contrapositive form to show that a series does not converge because the nth term does not approach zero. Note, however, that the *converse* of the theorem is not true as the following example will show. The above theorem is equivalent to the statement that $l_1 \subset (c_0)$. Note again, that $(c_0) \not\subset l_1$.

2.3.8 *EXAMPLE*   The harmonic series $\sum_{k=1}^{\infty} 1/k$ diverges. For, writing out the first few terms of the series shows that

$$A_1 = 1 = \frac{2}{2}, \quad A_2 = \frac{3}{2}, \quad A_4 > \frac{4}{2}, \quad A_8 > \frac{5}{2}, \quad A_{16} > \frac{6}{2}, \dots.$$

Then, $\lim_{n \to \infty} A_n = \infty$.

2.3.9 *THEOREM*   Let $\sum_{k=1}^{\infty} a_k$ and $\sum_{k=1}^{\infty} b_k$ be series of non-negative terms with $a_k \leq b_k$ for all k greater than some N. Then

(a) if $\sum_{k=1}^{\infty} b_k$ converges, $\sum_{k=1}^{\infty} a_k$ converges,
(b) if $\sum_{k=1}^{\infty} a_k$ diverges, $\sum_{k=1}^{\infty} b_k$ diverges.

Proof   The proof is left as problem 1.   ∎

2.3.10   *THEOREM*   Let $\sum_{k=1}^{\infty} a_k$ be a series of nonnegative terms. Let f be a continuous, positive, nonincreasing function defined for all $x \geq 1$, and let $a_k = f(k)$ for all k greater than some N. Then, $\sum_{k=1}^{\infty} a_k$ converges or diverges according as $\int_N^{\infty} f(x)\, dx$ converges or diverges.

Proof   The proof is left as Problem 2.   ∎

2.3.11   *THEOREM*   Let $\sum_{k=1}^{\infty} a_k$ be a series of positive terms and let

$$\lim_{n \to \infty} \frac{a_{n+1}}{a_n} = r.$$

Then,

(1) if $r < 1$, the series converges,
(2) if $r > 1$, the series diverges, and
(3) if $r = 1$, the series may converge or diverge.

Proof   The proof is left as Problem 3.   ∎

2.3.12   *THEOREM*   Let

$$\sum_{k=1}^{\infty} \frac{1}{k^p} = 1 + \frac{1}{2^p} + \frac{1}{3^p} + \cdots + \frac{1}{n^p} + \cdots,$$

where p is a constant. This is called the "p-series." Then,

(a) if $p > 1$, the series converges,

(b) if $p \leq 1$, the series diverges.

Proof   The proof is left as Problem 4.   ∎

2.3.13   *EXAMPLE*   To find $\lim_{n \to \infty} c^n/n!$, where $c > 0$, form the series $\sum_{k=1}^{\infty} c^k/k!$. By Theorem 2.3.11, the series converges, so by Theorem 2.3.7, $\lim_{n \to \infty} c^n/n! = 0$.

PROBLEMS†

1.   Prove Theorem 2.3.9.

2.   Prove Theorem 2.3.10.

3.   Prove Theorem 2.3.11.

4.   Prove Theorem 2.3.12.

5.   Use Theorem 2.3.9 to show that

$$l_1 \subset l_2 \subset l_3 \subset \cdots.$$

6.   Restate Theorem 2.3.12 in terms of $\{1/n\} \in l_p$.

7.   (a) Show geometrically that

$$\frac{1}{k} > \int_k^{k+1} \frac{dx}{x} > \frac{1}{k+1}$$

and hence

$$\frac{1}{k} - \frac{1}{k+1} > \frac{1}{k} - \int_k^{k+1} \frac{dx}{x} > 0,$$

for every $k = 1, 2, \ldots, n, \ldots$.

(b) Set

$$\frac{1}{k} - \int_k^{k+1} \frac{dx}{x} = c_k$$

and show $\lim_{n \to \infty} \sum_{k=1}^{n-1} c_k$ exists (finite).

† See also the problems following Section 2.4.

(c) Find $c_1 + c_2 + c_3 + \cdots + c_{n-1} = \sum_{k=1}^{n-1} c_k$ in closed form.

(d) Show

$$\lim_{n \to \infty} \left[\left(1 + \frac{1}{2} + \frac{1}{3} + \cdots + \frac{1}{n} \right) - \ln n \right] = \lim_{n \to \infty} \sum_{k=1}^{n-1} c_k.$$

This number is called *Euler's constant* or *Mascheroni's constant*. Its value to six decimal places is 0.577216. It is not known if it is transcendental or even if it is irrational.

8. The proof of the integral test (Theorem 2.3.10) gives a convenient way of estimating the difference between the partial sums and the sum of an infinite series. Prove the following: If the series $\sum_{k=1}^{\infty} a_k$ converges by the integral test [where $f(k) = a_k$], and if $A = A_n + R_n$, where A is the sum of the series and A_n the nth partial sum, then

$$| R_n | \leq \int_{n}^{\infty} f(x) \, dx.$$

9. Use Problem 8 to find R_6 if

$$A = \sum_{k=1}^{\infty} \frac{1}{k^4}.$$

10. How many terms of the series

$$\sum_{k=1}^{\infty} \frac{1}{k^2}$$

should be taken to give an approximation with an error less than 0.5?

11. How many terms of the series

$$\sum_{k=1}^{\infty} \frac{1}{k^2 + 1}$$

should be taken to give an approximation with an error less than 0.1?

12. The proof of the ratio test (Theorem 2.3.11) also suggests a method of estimating R_n. With the notation the same as in

Problem 8, prove the following: let $a_n > 0$, and let $\sum_{k=1}^{\infty} a_k$ converge by the ratio test. Then, if

$$\frac{a_{n+1}}{a_n} \le r < 1 \qquad \text{for} \quad n > N,$$

$$|R_n| \le \frac{a_{n+1}}{1 - r} \qquad \text{for} \quad n > N.$$

13.   Use Problem 12 to find R_6 if

$$A = \sum_{k=1}^{\infty} \frac{k+1}{k}\frac{1}{3^k}.$$

14.   How many terms of the series

$$\sum_{k=1}^{\infty} \frac{2^k + 1}{3^k}$$

should be taken to give an approximation with an error less than 0.5?

15.   In Problems 12–14 it was assumed not only that $\lim_{n\to\infty} a_{n+1}/a_n = r < 1$, but that $a_{n+1}/a_n \le r$. Devise a test in case $\lim_{n\to\infty} a_{n+1}/a_n = r < 1$, but $a_{n+1}/a_n > r$. (*Hint:*   consider $|a_{n+2}/a_{n+1}|$.)

2.4 additional convergence tests—alternating series

In addition to the tests for convergence given in Theorems 2.3.9–2.3.12, many other tests are known. The "big O" and the root tests can often be used in place of the comparison test and the ratio test, respectively. They are not a great deal more powerful but in some cases are easier to use.

2.4.1   *DEFINITION*   Let $\{a_n\}$ and $\{b_n\}$ be sequences of positive terms, and let $\lim_{n\to\infty} a_n/b_n = L$, where L may be infinite.

Then,

(1) If $0 \leq L < \infty$, $a_n = O(b_n)$.
(2) If $0 < L \leq \infty$, $b_n = O(a_n)$.
(3) If $0 < L < \infty$, $\{a_n\}$ and $\{b_n\}$ are of the same order of magnitude; that is $a_n = O(b_n)$ and $b_n = O(a_n)$.

2.4.2 THEOREM

(1) If $a_n = O(b_n)$ and $\sum_{k=1}^{\infty} b_k$ converges, $\sum_{k=1}^{\infty} a_k$ converges.
(2) If $b_n = O(a_n)$ and $\sum_{k=1}^{\infty} b_k$ diverges, $\sum_{k=1}^{\infty} a_k$ diverges.
(3) If $\{a_n\}$ and $\{b_n\}$ are of the same order of magnitude, the series $\sum_{k=1}^{\infty} a_k$ and $\sum_{k=1}^{\infty} b_k$ both converge or both diverge.

Proof We give the proof for (1) and leave the others as problems.

Since $\lim_{n \to \infty} a_n/b_n = L$, for any $\epsilon > 0$ there is an N such that if $n > N$, $a_n/b_n < L + \epsilon$. Thus $a_n < (L + \epsilon)b_n$. Then

$$\sum_{k=N+1}^{\infty} a_k < (L + \epsilon) \sum_{k=N+1}^{\infty} b_k.$$

As in the proof of Theorem 2.3.9, the convergence of $\sum_{k=1}^{\infty} b_k$ then implies convergence of $\sum_{k=1}^{\infty} a_k$. ∎

2.4.3 EXAMPLE Let

$$a_n = \frac{n}{n^2 + 3n + 1}, \qquad b_n = \frac{1}{n^2 + 2n - 1}, \qquad c_n = \frac{1}{\ln(n)},$$

$$d_n = \frac{2^n + n^2}{5^n}.$$

Then, $\{a_n\}$ is of the same order of magnitude as $\{1/n\}$, $\{b_n\}$ is of the same order of magnitude as $\{1/n^2\}$, $1/n = O(c_n)$, and $d_n = O(1/2^n)$

2.4.4 THEOREM Let $\sum_{k=1}^{\infty} a_k$ be a series of nonnegative terms, and let $\lim_{n \to \infty} (\sqrt[n]{a_n}) = s$, where $0 \leq s \leq \infty$.

(a) If $0 \le s < 1$, the series converges.
(b) If $1 < s \le \infty$, the series diverges.
(c) If $s = 1$, the series may converge or diverge.

Proof   We shall prove (a) only; the proof of (b) and the construction of two examples for (c) are left as problems. Since $\lim_{n\to\infty} (\sqrt[n]{a_n}) = s < 1$, let $\epsilon = (1 - s)/2$. For this ϵ there exists an N such that if $n > N$, we have

$$\sqrt[n]{a_n} < s + \frac{1 - s}{2} = \frac{s + 1}{2} < 1.$$

Let $t = (s + 1)/2$. Then, $a_n < t^n$, and $\sum_{k=N}^{\infty} a_k < \sum_{k=N}^{\infty} t^k$ with $t < 1$. Therefore, since the geometric series $\sum_{k=1}^{\infty} t^k$ converges, $\sum_{k=1}^{\infty} a_k$ converges.   ∎

The similarity between the ratio test and the root test is obvious. It can be proved (see Problems 5 and 6) that $r = 1$ in the ratio test if and only if $s = 1$ in the root test, so the advantages of one test over the other lie in the degree of difficulty in finding r and s, in the possibility that one limit exists and the other does not, and in the refinements that can be made.

Note that in the proof above we used the fact that

$$\sqrt[n]{a_n} < s + \frac{1 - s}{2}$$

but did not use the fact that

$$s - \frac{1 - s}{2} < \sqrt[n]{a_n}.$$

Thus, for convergence all we really need to have is

$$\limsup_{n\to\infty} \sqrt[n]{a_n} = s < 1.$$

Similarly, for divergence we need only have $\limsup_{n\to\infty} \sqrt[n]{a_n} = s > 1$. An examination of the proof of the ratio test reveals a similar situation except that lim inf is needed in the latter case. (See Problems 3 and 4.)

2.4.5 *EXAMPLE*   Let

$$
a_n = \begin{cases}
\dfrac{1}{2^{(n+1)/2}} & \text{if } n \text{ is odd,} \\[2ex]
\dfrac{1}{3^{n/2}} & \text{if } n \text{ is even.}
\end{cases}
$$

Then

$$
\limsup_{n \to \infty} \frac{a_{n+1}}{a_n} = +\infty, \qquad \liminf_{n \to \infty} \frac{a_{n+1}}{a_n} = 0,
$$

so even in its refined form the ratio test fails. But

$$
\limsup_{n \to \infty} \sqrt[n]{a_n} = \frac{1}{\sqrt{2}},
$$

so the refined form of the root test shows convergence of the series.

We have dealt thus far primarily with series whose terms are non-negative. Series whose terms may be positive or negative may offer special problems, but one important theorem is true for all such series.

2.4.6   *THEOREM*   If $\sum_{k=1}^{\infty} a_k$, where a_k may be positive, negative, or zero, is absolutely convergent, $\sum_{k=1}^{\infty} a_k$ is convergent.

Proof   Let $A_n = \sum_{k=1}^{n} a_k$, $b_k = |a_k|$, $B_n = \sum_{k=1}^{n} b_k$, $c_k = a_k + |a_k|$, $C_n = \sum_{k=1}^{n} c_k$, and $B = \lim_{n \to \infty} B_n$ (which exists by hypothesis). Then, since $c_k = 2|a_k|$ or 0, $0 \le C_n \le 2B_n \le 2B$, and $\{C_n\}$ is nondecreasing. Therefore, $\lim_{n \to \infty} C_n$ exists. Call it C. Then, $\lim_{n \to \infty} A_n = C - B$. ∎

The following theorem applies to *strictly* alternating series:

2.4.7   *THEOREM*   If a series satisfies the following conditions:

(a) it is strictly alternating, that is, $a_n > 0$ and the signs follow the pattern $a_1 - a_2 + a_3 - a_4 + \cdots$,

(b) $a_{n+1} \le a_n$, $n = 1, 2, \ldots$,

(c) $\lim_{n \to \infty} a_n = 0$,

then $\sum_{k=1}^{\infty} (-1)^{k+1} a_k$ converges.

Proof   Let

$$A_{2n} = (a_1 - a_2) + (a_3 - a_4) + \cdots + (a_{2n-1} - a_{2n}),$$

and

$$A_{2n+1} = a_1 - (a_2 - a_3) - (a_4 - a_5) - \cdots - (a_{2n} - a_{2n+1}).$$

Then A_{2n} is monotone nondecreasing and A_{2n+1} is monotone nonincreasing. Since A_{2n} is bounded above and A_{2n+1} is bounded below, they both have limits as $n \to \infty$ by Theorem 2.1.4; and since $\lim_{n \to \infty} a_{2n+1} = 0$, these limits must be the same, say, A. Then, $\lim_{n \to \infty} A_n = A$.   ∎

2.4.8 *COROLLARY*   With the hypotheses stated above, let $A = A_n + R_n$ with $\lim_{n \to \infty} A_n = A$. Then $|R_n| < a_{n+1}$.

Proof   If n is odd, let $n = 2k - 1$. Then

$$0 < A_{2k-1} - A < A_{2k-1} - A_{2k} = a_{2k}.$$

If n is even, let $n = 2k$. Then

$$0 < A - A_{2k} < A_{2k+1} - A_{2k} = a_{2k+1}.$$ ∎

2.4.9 *DEFINITION*   If an alternating series converges but does not converge absolutely, it is said to *converge conditionally*.

2.4.10 *EXAMPLE*   The series

$$\sum_{k=1}^{\infty} (-1)^{k+1} \left(\frac{1}{k} \right)$$

converges but does not converge absolutely. Therefore, it converges conditionally.

Problems

1. Complete the proof of Theorem 2.4.2.

2. Complete the proof of Theorem 2.4.4.

3. Prove the following form of the ratio test: let $a_n > 0$. Then $\sum_{n=1}^{\infty} a_n$ converges if

$$\limsup_{n \to \infty} \frac{a_{n+1}}{a_n} = r < 1$$

and diverges if

$$\liminf_{n \to \infty} \frac{a_{n+1}}{a_n} = r > 1.$$

4. Prove the following form of the root test: let $a_n > 0$. Let $s = \limsup_{n \to \infty} \sqrt[n]{a_n} = s$. Then, $\sum_{n=1}^{\infty} a_n$ converges if $s < 1$ and diverges if $s > 1$.

5. Let $\sum_{k=1}^{\infty} a_k$ be a series of nonnegative terms. Prove that

$$\liminf_{n \to \infty} \frac{a_{n+1}}{a_n} \leq \liminf_{n \to \infty} \sqrt[n]{a_n} \leq \limsup_{n \to \infty} \sqrt[n]{a_n} \leq \limsup_{n \to \infty} \frac{a_{n+1}}{a_n}.$$

6. Using the results in Problem 5, state any conclusions that can be drawn concerning examples where the root test succeeds or fails, the ratio test succeeds or fails, and their refinements succeed or fail.

7. From Problems 5 and 6, prove

$$\lim_{n \to \infty} \frac{n}{\sqrt[n]{n!}} = e.$$

In Problems 8–34, determine convergence or divergence of the given series. If an alternating series converges, test for absolute or conditional convergence.

8. $\displaystyle\sum_{n=1}^{\infty} \frac{1}{\sqrt{n^2 + 1}}.$ **9.** $\displaystyle\sum_{n=1}^{\infty} \frac{n - 1}{n^3}.$

10. $\displaystyle\sum_{n=1}^{\infty} \frac{n^2}{e^n}.$

11. $\displaystyle\sum_{n=1}^{\infty} \frac{(-2)^n}{3^n + 1}.$

12. $\displaystyle\sum_{n=1}^{\infty} \frac{\ln n}{n^2}.$

13. $\displaystyle\sum_{n=1}^{\infty} \frac{e^n}{n^n}.$

14. $\displaystyle\sum_{n=2}^{\infty} \frac{1}{n \ln(n)}.$

15. $\displaystyle\sum_{n=2}^{\infty} \frac{1}{n \ln^2(n)}.$

16. $\displaystyle\sum_{n=1}^{\infty} \frac{\sqrt{n^2 + 1}}{n^3 + 1}.$

17. $\displaystyle\sum_{n=1}^{\infty} \frac{(-1)^n n!}{3^n}.$

18. $\displaystyle\sum_{n=1}^{\infty} a_n,$ where

$$a_n = \begin{cases} \dfrac{-1}{3^{(n+1)/2}} & \text{if } n \text{ is odd,} \\[2em] \dfrac{+1}{2^{n/2}} & \text{if } n \text{ is even.} \end{cases}$$

19. $\displaystyle\sum_{n=1}^{\infty} \frac{1\cdot 2\cdot 3\cdots n}{1\cdot 3\cdot 5\cdots(2n-1)}.$

20. $\displaystyle\sum_{n=1}^{\infty} \frac{(-1)^n n}{n^2 - \pi^2}.$

21. $\displaystyle\sum_{n=1}^{\infty} \frac{(-1)^n \ln(n)}{n}.$

22. $\displaystyle\sum_{n=2}^{\infty} \frac{\ln(n+1) - \ln(n-1)}{n}.$

23. $\displaystyle\sum_{n=1}^{\infty} \frac{\ln n}{n^2 + 1}.$

24. $\displaystyle\sum_{n=1}^{\infty} n^2 e^{-n}.$

25. $\displaystyle\sum_{n=1}^{\infty} \frac{n!}{n^n}.$

26. $\displaystyle\sum_{n=1}^{\infty} \frac{(-1)^n}{n + \dfrac{1}{n}}.$

27. $\displaystyle\sum_{n=1}^{\infty} \frac{(-1)^n}{n^{1+(1/n)}}.$

28. $\displaystyle\sum_{n=1}^{\infty} \frac{1}{\left(1 + \dfrac{1}{n}\right)^{n^2}}.$

29. $\displaystyle\sum_{n=2}^{\infty} \frac{1}{(\ln n)^n}.$

30. $\displaystyle\sum_{n=1}^{\infty} \frac{\cos n\pi}{n + 1}.$

31. $\displaystyle\sum_{n=1}^{\infty}\left(\frac{2n+1}{3n}\right)^n.$ **33.** $\displaystyle\sum_{n=1}^{\infty}\ln\left(1+\frac{1}{n}\right).$

32. $\displaystyle\sum_{n=1}^{\infty}\frac{n!}{3\cdot5\cdot7\cdots(2n+3)}.$ **34.** $\displaystyle\sum_{n=1}^{\infty}\ln\left(1+\frac{1}{n^2}\right).$

35. Generalize the results of Problems 33 and 34. That is, state and prove a theorem concerning convergence or divergence of $\sum_{n=1}^{\infty}\ln(1+a_n)$.

36. If an infinite *product* is to converge, the limit of the nth factor must be 1, just as the limit of the nth term of a convergence *series* must be 0. Using the results of Problem 35, speculate on a necessary and sufficient condition for $\prod_{n=1}^{\infty}(1+a_n)$ to converge.

2.5 convergence sets for power series

In order to investigate power series thoroughly we need some properties of the derivative and integral not yet considered. We may, however, find the set of numbers for which a power series converges by methods of the preceding sections.

2.5.1 EXAMPLE Find all real numbers x for which the following power series converges.

$$\sum_{n=0}^{\infty}\frac{x^n}{(n+1)2^n}$$

Solution First consider absolute convergence and apply the ratio test. Let

$$r = \lim_{n\to\infty}\left|\frac{x^{n+1}/(n+2)2^{n+1}}{x^n/(n+1)2^n}\right|$$

$$= \lim_{n\to\infty}\frac{|x|}{2}\frac{(n+1)}{(n+2)} = \frac{|x|}{2}.$$

Now, set $|x|/2 < 1$. The series converges absolutely for $-2 < x < 2$ and diverges if $x > 2$ or $x < -2$. For $x = 2$ or -2, the ratio tests fails, but the "big O" test shows divergence for $x = 2$ and Theorem 2.4.7 shows conditional convergence for $x = -2$.

2.5.2   EXAMPLES   The ratio test shows that the series $\sum_{n=1}^{\infty} n! \, x_n$ diverges for $x \neq 0$ and that the series $\sum_{n=1}^{\infty} x^n/n!$ converges for all x.

If a power series converges for $|x| < r$ and diverges for $|x| > r$, the interval $(-r, +r)$ is called the *interval of convergence*, and r is the radius of convergence without regard to convergence or divergence at $x = \pm r$. (It is not customary to say the interval of convergence is closed to indicate convergence at the end points.)

The series considered in this section have all been power series in x, that is, they are of the form

$$\sum_{n=0}^{\infty} a_n x^n.$$

We can, of course, have a power series in $(x - a)$ as

$$\sum_{n=0}^{\infty} a_n (x - a)^n.$$

We may then have convergence for $|x - a| < r$ and divergence for $|x - a| > r$. Again, r is the radius of convergence, and the interval of convergence is $(a - r, a + r)$.

PROBLEMS

Find the interval of convergence and determine convergence or divergence at the end points for Problems 1–12.

1. $\displaystyle\sum_{n=0}^{\infty} \frac{n(x - 2)^n}{2^n (n^2 + 1)}.$

2. $\displaystyle\sum_{n=1}^{\infty} \frac{x^{2n-1}}{(2n - 1)!}.$

3.† $\displaystyle\sum_{n=1}^{\infty} \frac{x^{n-1}n!}{n^n}.$

8. $\displaystyle\sum_{n=1}^{\infty} \frac{1}{x^n - 1}.$

4. $\displaystyle\sum_{n=1}^{\infty} \frac{(nx)^{n-1}}{n!}.$

9. $\displaystyle\sum_{n=1}^{\infty} \frac{e^{nx}}{2^n}.$

5. $\displaystyle\sum_{n=0}^{\infty} \frac{(x+3)^n}{4^n(n+1)^{1/2}}.$

10. $\displaystyle\sum_{n=1}^{\infty} \frac{x^n}{(1-x)^n}.$

6. $\displaystyle\sum_{n=1}^{\infty} \frac{(nx)^n}{n^3+1}.$

11. $\displaystyle\sum_{n=0}^{\infty} \frac{1}{(1+x)^n}.$

7. $\displaystyle\sum_{n=1}^{\infty} n^2 e^{-nx}.$

12. $\displaystyle\sum_{n=0}^{\infty} e^{nx}.$

13. In Problems 11 and 12, find the sum.

2.6 operations on series

We have noted that l_p $(1 \leq p < \infty)$ is a normed linear space (see Problem 6, Section 2.2) which requires, among other things, that if $x_1 \in l_p$ and $x_2 \in l_p$, then $x_1 + x_2 \in l_p$. If $p = 1$, this means that the sum of two absolutely convergent series is absolutely convergent. We shall see (Problem 1) that the condition of absolute convergence can be relaxed and that the sum of two (absolutely or conditionally) convergent series converges, provided the order of addition is not disturbed.

That the order of addition of elements in a conditionally convergent series is important can be seen by considering the series

† To test the end points for convergence in this and certain other problems it is convenient to use Stirling's formula which states

$$\lim_{n \to \infty} \frac{n!}{(n/e)^n \sqrt{2\pi n}} = 1.$$

For a proof, see Taylor, A. E., "Advanced Calculus," Section 20.8. Ginn (Blaisdell), Boston, Massachusetts, 1955.

$1 - \frac{1}{2} + \frac{1}{3} - \frac{1}{4} + \frac{1}{5} - \cdots$, which gives rise to two divergent series, $1 + \frac{1}{3} + \frac{1}{5} + \cdots$, and $-(\frac{1}{2} + \frac{1}{4} + \frac{1}{6} + \cdots)$. In such a case a rearrangement of the series may be made to converge to any prescribed number. For example, $1 + \frac{1}{3} + \frac{1}{5} + \cdots + \frac{1}{15} > 2$, but $1 + \frac{1}{3} + \cdots + \frac{1}{15} - \frac{1}{2} < 2$. Continuing, we may take more terms of the first series (since it diverges) to make the finite sum again greater than 2. Then add $-\frac{1}{4}$. Thus we obtain a series whose partial sums converge to 2. This cannot happen for an absolutely convergent series, as we shall see.

2.6.1 *DEFINITION* Let Φ be a one-to-one mapping of the positive integers *onto* themselves; that is, if s, t, u, and v are positive integers, then for each s there exists a unique t such that $\Phi(s) = t$, and for each v there exists a unique u such that $\Phi(u) = v$. Then, $\{a_{\Phi(n)}\}$ is called a *rearrangement* of $\{a_n\}$. The set $\{a_{\Phi(n)}\} = \{a_{\Phi(1)}, a_{\Phi(2)}, \ldots\}$ is usually written $\{a_{k_1}, a_{k_2}, a_{k_3}, \ldots\} = \{a_{k_n}\}$.

2.6.2 *THEOREM* If $\sum_{k=1}^{\infty} a_k$ converges absolutely, then every rearrangement converges absolutely, and all rearrangements have the same sum.

Proof Let $\sum_{k=1}^{\infty} a_k = A$, $\sum_{k=1}^{\infty} |a_k| = B$ and let $\{a_{k_i}\}$ be a rearrangement of $\{a_k\}$. Then $\sum_{i=1}^{n} |a_{k_i}| \leq B$ for every n, so every rearrangement converges absolutely. To prove all sums are equal fix $\epsilon > 0$. There is an N such that

$$\left| A - \sum_{k=1}^{N} a_k \right| < \frac{\epsilon}{2} \quad \text{and} \quad \sum_{k=N+1}^{\infty} |a_k| < \frac{\epsilon}{2}.$$

Let N' be such that $a_1, a_2, \ldots, a_N$ are included in $a_{k_1}, a_{k_2}, \ldots, a_{k_{N'}}$. Then, if $n > N'$,

$$\left| A - \sum_{i=1}^{n} a_{k_i} \right| < \left| A - \sum_{i=1}^{N} a_i \right| + \sum_{k=N+1}^{\infty} |a_k|$$

$$< \frac{\epsilon}{2} + \frac{\epsilon}{2} = \epsilon.$$

Thus $A = \sum_{i=1}^{\infty} a_{k_i}$. ∎

The above theorem assumes great importance when we consider the problem of multiplying two infinite series together to obtain a new series. It is not surprising that many different arrangements of the product are used when we realize that there is little agreement on the order of terms in the product of polynomials, if they contain more than two terms. Our procedure is to show that *one* arrangement of the terms in the product of two absolutely convergent series converges, and then by the previous theorem any other arrangement will do as a formula.

2.6.3   *THEOREM*   If $A = \sum_{k=1}^{\infty} a_k$ and $B = \sum_{k=1}^{\infty} b_k$ are two absolutely convergent series, there is an arrangement of the products $a_i b_j$ which forms an absolutely convergent series $\sum_{k=1}^{\infty} c_k = C$, where the set of the c_k's is the same as the set of all $a_i b_j$, and $C = AB$.

Proof   Let $c_1 = a_1 b_1, c_2 = a_2 b_1, c_3 = a_2 b_2, c_4 = a_1 b_2, \ldots$, where c_k is the kth element in the array

$$
\begin{array}{c|c|c|c}
a_1 b_1 & a_1 b_2 & a_1 b_3 & \cdots \\
\hline
a_2 b_1 & a_2 b_2 & a_2 b_3 & \cdots \\
\hline
a_3 b_1 & a_3 b_2 & a_3 b_3 & \cdots \\
\vdots & \vdots & \vdots
\end{array}
$$

found by counting in a counterclockwise direction around $a_1 b_1$ in the paths outlined. Then,

$$\sum_{k=1}^{n^2} |c_k| = \left(\sum_{k=1}^{n} |a_k|\right)\left(\sum_{k=1}^{n} |b_k|\right)$$

$$\leq \left(\sum_{k=1}^{\infty} |a_k|\right)\left(\sum_{k=1}^{\infty} |b_k|\right),$$

so, $\sum_{k=1}^{\infty} c_k$ is absolutely convergent.   ∎

2.6.4   *COROLLARY*   If $\sum_{k=1}^{\infty} a_k$ and $\sum_{k=1}^{\infty} b_k$ are absolutely convergent series, and if $c_k = a_i b_j$, where $i = 1, 2, \ldots,$

$j = 1, 2, \ldots$, then $\sum_{k=1}^{\infty} c_k$ is absolutely convergent for *any* arrangement of $a_i b_j$ with each a_i and each b_j used once and only once.

Proof This follows directly from Theorems 2.6.3 and 2.6.2. ∎

2.6.5 *REMARK* The Cauchy product

$$a_1 b_1 + (a_2 b_1 + a_1 b_2) + (a_3 b_1 + a_2 b_2 + a_1 b_3)$$
$$+ (a_4 b_1 + a_3 b_2 + a_2 b_3 + a_1 b_4) + \cdots$$

is an easy one to remember. It is formed by taking diagonals in the array in the proof of Theorem 2.6.3.

Problems

1. If $\sum_{k=1}^{\infty} a_k$ and $\sum_{k=1}^{\infty} b_k$ are convergent series, prove that $\sum_{k=1}^{\infty} (a_k + b_k)$ converges to $\sum_{k=1}^{\infty} a_k + \sum_{k=1}^{\infty} b_k$.

2. Let
$$k = \frac{(i + j - 2)(i + j - 1)}{2} + j.$$

Verify that $c_k = a_i b_j$ gives the Cauchy product for the above formula for k.

3. Find several partial sums of a rearrangement of $1 - \frac{1}{2} + \frac{1}{3} - \frac{1}{4} + \cdots$ that are alternately greater than and less than 10. (A computer is helpful!)

4. Prove the results in Corollary 2.6.4 assuming only one of the series is absolutely convergent and the other just convergent.

5. Criticize the following "proof": let

$$S = 1 - \tfrac{1}{2} + \tfrac{1}{3} - \tfrac{1}{4} + \tfrac{1}{5} - \tfrac{1}{6} + \tfrac{1}{7} - \tfrac{1}{8} + \tfrac{1}{9} - \cdots$$

$$\tfrac{1}{2}S = \qquad \tfrac{1}{2} \qquad - \tfrac{1}{4} \qquad + \tfrac{1}{6} \qquad - \tfrac{1}{8} \qquad + \cdots$$

$$\tfrac{3}{2}S = 1 \qquad + \tfrac{1}{3} - \tfrac{1}{2} + \tfrac{1}{5} \qquad + \tfrac{1}{7} - \tfrac{1}{4} + \tfrac{1}{9} + \cdots$$

$$= S.$$

6. Let

$$S = 1 - \frac{1}{\sqrt{2}} + \frac{1}{\sqrt{3}} - \frac{1}{\sqrt{4}} + \frac{1}{\sqrt{5}} - \cdots$$

and form the Cauchy product of S by itself. Show that the resulting series diverges.

7. Verify that the proofs of Hölder's inequality and Minkowski's inequality given in Section 1.3 hold for infinite sums.

8. Under what conditions does equality hold in each part of Problem 7.

9. If $\sum_{k=1}^{\infty} a_k$ and $\sum_{k=1}^{\infty} b_k$ are convergent series of positive real numbers, prove that $\sum_{k=1}^{\infty} a_k^{1/2} b_k^{1/2}$ converges.

10. If $\sum_{k=1}^{\infty} a_k^2$ and $\sum_{k=1}^{\infty} b_k^2$ converge, show that $\sum_{k=1}^{\infty} (a_k + b_k)^2$ converges.

III

COMPLETENESS PROPERTIES

3.1 completeness of the real number system

A rational number can be written as the quotient of two integers p/q, where $q \neq 0$.

The decimal expansion of a real number is actually a geometric series, finite or infinite, whose terms are of the form $a_k 10^k$, where $k = 0, \pm 1, \pm 2, \ldots$, Thus, for example, $235.74 = 2 \cdot 10^2 + 3 \cdot 10^1 + 5 \cdot 10^0 + 7 \cdot 10^{-1} + 4 \cdot 10^{-2}$, and $0.3333\ldots = 3 \cdot 10^{-1} + 3 \cdot 10^{-2} + \cdots$. It is obvious, then, that any terminating decimal is rational; and a repeating decimal may be treated as an infinite series or may be shown to be rational as in the following example. (In order to have a *unique* decimal representation an agreement must be made concerning an infinite repetition of 9's. Since $0.999 \ldots$ is an infinite geometric series whose sum is $1.000 \ldots$, we will agree to replace an infinite repetition of 9's by 0's and increase the preceding digit by 1.)

3.1.1 *EXAMPLE* Let

$$x = 2.5\ 81\ 81\ 81\ldots.$$

Then

$$10x = \quad 25.81\ 81\ 81\ldots,$$

$$1000x = 2581.81\ 81\ 81\ldots,$$

$$990x = 2556,$$

$$x = \frac{2556}{990} = \frac{142}{55}.$$

Let us construct a sequence of rational numbers from the familiar algorithm for $\sqrt{2}$. Thus, $a_1 = 1$, $a_2 = 1.4$, $a_3 = 1.41$, $a_4 = 1.414$, $a_5 = 1.4142$, ... This sequence of rational numbers converges to $\sqrt{2}$, but $\sqrt{2}$ is not rational as a simple proof demonstrates (see Problem 1).

Thus the rational numbers do not provide a rich enough set of numbers for our needs. We are convinced that $\sqrt{2}$ is a *real number* (we know its position on the real axis as Fig. 1 shows), it is the limit of a sequence of rational numbers, but it is not itself rational. This unfortunate state of affairs is expressed by saying that the rational numbers are not *complete*. If we arrange the rational numbers and only the rationals numbers on a straight line according to their natural order, there is a "hole" where $\sqrt{2}$ "should be." Two questions naturally arise:

(1) How can we obtain all real numbers from the simpler system of rational numbers?

(2) Are the real numbers thus obtained themselves complete?

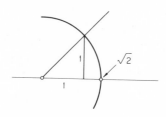

Figure 1

A detailed discussion of these questions would require considerable digression, and the student is referred to the work of Landau.[†] For our purposes, we shall assume the familiar algebraic properties of real numbers, a completeness axiom, and also the properties of the order relations "$<$" and "$\leq$".[‡] Among the last is the property that between any two numbers (rational or irrational) there is both a rational and an irrational number (and therefore an infinite number of each kind).

The property of real numbers that concerns us most at present is that of completeness. It will be seen that there are many precise ways of describing this intuitive property, but the following axiom is probably the simplest and most direct approach.

3.1.2 COMPLETENESS AXIOM In the real number system every nonempty set of numbers bounded above has a least upper bound, or supremum.

It can then be seen that every nonempty set bounded below has a greatest lower bound, or infimum.

Referring to our example above, there exists a real number a such that $a^2 = 2$; 1.4, 1.41, 1.414, $\ldots \leq a$; and, for every $\epsilon > 0$, there are numbers in the sequence greater than $a - \epsilon$. Then a is written $\sqrt{2}$.

We are now in a position to prove Theorem 2.1.4 which is restated here for convenience:

3.1.3 THEOREM Every bounded monotone sequence of real numbers converges.

> **Proof** We prove this for the case that $\{a_n\}$ is nondecreasing and leave the others as problems. Since $\{a_n\}$ is bounded above, there exists a supremum, call it a. Then, $a_n \leq a$ for every n. Also, for $\epsilon > 0$, there is at least one number in the

---

† Landau, E. "Foundations of Analysis" 2nd ed. Chelsea, Bronx, New York, 1960.

‡ Such a system is referred to as a *complete ordered field*.

sequence, call it a_N, such that $a_N > a - \epsilon$. Because of the monotonic nature of $\{a_n\}$, $a_n > a - \epsilon$ if $n > N$. Obviously, $a_n < a + \epsilon$, so $|a_n - a| < \epsilon$ if $n \geq N$. That is, $\lim_{n \to \infty} a_n = a$. ∎

Returning briefly to the rational numbers, a somewhat surprising feature is that, in a sense, there are no more rational numbers than there are positive integers. The following definition and theorem make this precise:

3.1.4   *DEFINITION*   A set of elements that can be put into a one-to-one correspondence with the positive integers is said to *countable*, or *denumerable*.

3.1.5   *THEOREM*   The rational numbers are countable.

Proof   Arrange the positive rationals as

$$1, \quad 2, \quad 3, \quad \ldots$$
$$\tfrac{1}{2}, \quad \tfrac{2}{2}, \quad \tfrac{3}{2}, \quad \ldots$$
$$\tfrac{1}{3}, \quad \tfrac{2}{3}, \quad \tfrac{3}{3}, \quad \ldots$$

Figure 2

and count them as in Fig. 2, omitting those already counted in another form.

Then count the negative rationals in the same way and finally count them together as

$$p_1 \to p_2 \quad p_3 \to p_4 \quad \cdots$$
$$n_1 \to n_2 \quad n_3 \to n_4 \quad \cdots. \quad ∎$$

The completeness axiom for real numbers depends on an order relation ("$<$" or "$\leq$") which is one of the features of real numbers but is not present in all linear spaces. In order to extend the idea of completeness to other linear spaces a different approach must be made. Note that in our example, numbers in the sequence 1.4, 1.41, 1.414, . . . are not only getting close to *something*, $\sqrt{2}$, but are getting close to *each other*. This turns out to be a clue for defining completeness is other spaces, and the following definition expresses the idea exactly:

3.1.6 DEFINITION The sequence $\{a_n\}$ is a Cauchy sequence if

$$\lim_{m,n \to \infty} |a_n - a_m| = 0;$$

that is, if for every $\epsilon > 0$, there is an N such that if $n > N$ and $m > N$, $|a_n - a_m| < \epsilon$.

This can obviously be extended to general normed linear spaces if absolute values are replaced by norms.

We shall see that for the real number system the completeness axiom is equivalent to the property that every Cauchy sequence converges to a real number. Unfortunately, this is not true of every system that has both an order and a distance concept; but, for the linear spaces we will consider, this difficulty does not arise. First, a trivial theorem:

3.1.7 THEOREM If $\lim_{n \to \infty} a_n = a$, $\{a_n\}$ is a Cauchy sequence.

Proof Fix $\epsilon > 0$. Then there exists an N such that if $n > N$ and $m > N$,

$$|a_n - a| < \frac{\epsilon}{2} \quad \text{and} \quad |a_m - a| < \frac{\epsilon}{2}.$$

Then,

$$|a_n - a_m| \leq |a_n - a| + |a - a_m| < \frac{\epsilon}{2} + \frac{\epsilon}{2} = \epsilon. \quad \blacksquare$$

3.1.8 THEOREM (a) Assuming the completeness axiom, every Cauchy sequence of real numbers converges to a real number.

(b)   Assuming that every Cauchy sequence of real numbers converges to a real number, the completeness axiom holds.

Proof of (a)   Let $\{a_n\}$ be a Cauchy sequence. Then $\{a_n\}$ is bounded above. For, note that for a fixed ϵ, say, $\epsilon = 1$, there is an N such that if m, $n > N$, $|a_n - a_m| < 1$; in particular, $|a_n - a_{N+1}| < 1$ if $n > N$. Now, take $M = \max(a_1, a_2, \ldots, a_N, a_{N+1} + 1)$. Then, $a_n \leq M$ for all n. Similarly, $\{a_n\}$ is bounded below.

Now, assuming the completeness axiom, there is $A_k = \inf\{a_n\}$ for $n \geq k$. Then $\{A_k\}$ is bounded and monotone nondecreasing. So, let $A = \lim_{k \to \infty} A_k$. Now, fix $\epsilon > 0$. There exists an N' such that $|a_n - a_m| < \epsilon/3$ if n, $m > N'$, and an N'' such that $|A_k - A| < \epsilon/3$ if $k > N''$. Let $N = \max(N', N'')$; so for n, $m > N$ and $k = N + 1$, $|a_n - a_m| < \epsilon/3$ and $|A_{N+1} - A| < \epsilon/3$. By the definition of A_{N+1}, there is a k' $(k' \geq N + 1)$ such that $a_{k'} - A_{N+1} < \epsilon/3$. Now, if $n > N$,

$$|A - a_n| \leq |A - A_{N+1}| + |A_{N+1} - a_{k'}| + |a_{k'} - a_n|$$

$$< \frac{\epsilon}{3} + \frac{\epsilon}{3} + \frac{\epsilon}{3} = \epsilon. \quad \blacksquare$$

Proof of (b)   Let S be a set of real numbers bounded above. Without loss of generality we may assume M is an upper bound and $M - 1$ is not. Then determine whether $M - \frac{1}{2}$ is or is not an upper bound and choose the interval $[M - 1, M - \frac{1}{2}]$ if it is or $[M - \frac{1}{2}, M]$ if it is not. Similarly, choose $[M - 1, M - \frac{3}{4}]$, $[M - \frac{3}{4}, M - \frac{1}{2}]$, $[M - \frac{1}{2}, M - \frac{1}{4}]$, or $[M - \frac{1}{4}, M]$ so that the right-hand end point is the smallest upper bound among the numbers $M - \frac{3}{4}$, $M - \frac{1}{2}$, $M - \frac{1}{4}$, M. We thus have intervals of lengths 1, $\frac{1}{2}$, $(\frac{1}{2})^2$. Continue forming intervals of lengths $(\frac{1}{2})^n$ so that the right-hand end points are upper bounds of S and the left-hand end points are not. These right-hand end points form a Cauchy sequence which by assumption has a limit, say, a. Then $a = \sup S$. $\quad \blacksquare$

PROBLEMS

1. Prove that $\sqrt{2}$ is irrational. (*Hint*: assume $\sqrt{2} = p/q$, where p and q have no common factors. Square both sides and show that p and q both have a factor of 2 which is a contradiction.)

2. Prove that $\sqrt{3}$ is irrational.

3. Prove the following are irrational:
   (a) $2\sqrt{2} + 3$,
   (b) $\sqrt[3]{2}$,
   (c) $\sqrt{2} + \sqrt{3}$.

4. Express each of the following as a rational number:
   (a) $3.1428\ 428\ 428\ldots$.
   (b) $2.142857\ 142857\ 142857\ldots$.
   (c) $-1.27\ 27\ 27\ldots$.

5. Complete the proof of Theorem 3.1.3.

6. Prove the real numbers are not countable. [*Hint*: suppose they are. Then
$$x_1 = 0.a_{11}\ a_{12}\ a_{13}\ldots,$$
$$x_2 = 0.a_{21}\ a_{22}\ a_{23}\ldots,$$
$$x_3 = 0.a_{31}\ a_{32}\ a_{33}\ldots,$$
$$\vdots$$
where a_{ij} is the jth digit in the decimal expansion of the ith number (arranged for counting) between 0 and 1. Where is $0.\bar{a}_{11}\ \bar{a}_{22}\ \bar{a}_{33}\ldots$, where $\bar{a}_{kk} \neq a_{kk}$?]

7. Define a Cauchy sequence in a metric space. (See Problem 13, Section 1.2.)

8. Show directly from the definition that $\{1/n\}$ is a Cauchy sequence.

9. Let
$$x_1 = 1, \quad x_2 = \frac{1}{2}, \quad x_3 = \frac{3}{4}, \quad x_4 = \frac{5}{8}, \ldots, x_n = \frac{x_{n-2} + x_{n-1}}{2}.$$

Show directly from the definition that $\{x_n\}$ is a Cauchy sequence.

10.   Prove that the limit superior of a sequence always exists if $+\infty$ and $-\infty$ are allowed as limits.

11.   Theorem 3.1.8(a) is true for real numbers, but not necessarily true in a general linear space; that is, Cauchy sequences do not always converge *to an element in the space*. However, prove the following: suppose the Cauchy sequence $\{a_n\}$ has a subsequence[†] that converges to a. Prove $\{a_n\}$ converges to a.

3.2 norm convergence

The preceding section showed that for the real number system we have two equivalent conditions for completeness. In considering the completeness of linear spaces, however, we do not in general have an order relation, so the Cauchy criterion for convergence [Theorem 3.1.8(a)] will be used for the definition:

3.2.1   *DEFINITION*   A normed linear space is *complete* if and only if every Cauchy sequence converges to an element in the space.

The student has probably wondered about the seemingly arbitrary definitions of a norm in the linear spaces studied thus far, and it is now possible to justify our definitions. For, it turns out that *under the norms defined previously, the spaces $C[a, b]$, $B[a, b]$, $C^{(1)}[a, b]$, and all sequence spaces defined here are complete*. This property will be proved in each case when the space is studied in detail. Complete normed linear spaces assume great importance in modern mathematics and are called *Banach spaces*, after Stefan Banach (1892–1945) who is considered one of the founders of modern functional analysis.

When the norm $\| f \|$ is defined in terms of $\sup | f(x) |$, a Cauchy sequence of elements of the function space gives rise to a Cauchy sequence of real numbers. That is, if

$$\| f_n - f_m \| = \sup | f_n(x) - f_m(x) | \to 0,$$

---

† Let $\{a_n\} = \{a_1, a_2, a_3, \ldots\}$ be a sequence, and let $n_1 < n_2 < n_3 < \ldots$ be positive integers. Then, $\{a_{n_1}, a_{n_2}, a_{n_3}, \ldots\} = \{a_{n_k}\}$ is a subsequence of $\{a_n\}$.

we have

$$| f_n(x) - f_m(x) | \to 0 \qquad \text{for each} \quad x.$$

(Note that the converse is not true.) Completeness of the real numbers then guarantees the existence of a pointwise limit function f where $\lim_{n \to \infty} | f_n(x) - f(x) | = 0$. The problem of completeness in function spaces is thus generally reduced to showing that (1) pointwise convergence is also norm convergence and (2) the limit function is an element of the space.

3.2.2   *EXAMPLE*   Let $f_n(x) = x_n,\ 0 \le x \le 1$. Then, $f_n \in C[0, 1]$ and

$$\lim_{n \to \infty} f_n(x) = f(x) = \begin{cases} 0 & \text{if} \quad 0 \le x < 1, \\ 1 & \text{if} \quad x = 1, \end{cases}$$

and $f \notin C[0, 1]$. So, accepting the completeness of $C[0, 1]$, this pointwise limit cannot be the limit in the norm of $C[0, 1]$. To verify this directly, consider

$$\| f_n - f \| = \sup_{0 \le x \le 1} | f_n(x) - f(x) | = \sup_{0 \le x < 1} | f_n(x) - f(x) |$$

$$= \sup_{0 \le x < 1} x^n = 1.$$

Thus, $\lim_{n \to \infty} \| f_n - f \| = 1 \ne 0$, and $f_n \to f$ in the norm of $C[0, 1]$.

3.2.3   *EXAMPLE*   Let

$$f_n(x) = \frac{x}{1 + nx^2}, \qquad 0 \le x \le 1.$$

Elementary calculus gives the maximum value to be $f_n(1/\sqrt{n}) = 1/2\sqrt{n}$. Also, $f_n(x) \ge 0$ for all n, and $\lim_{n \to \infty} f_n(x) = f(x) = 0$. Then,

$$\| f_n - f \| = \sup | f_n(x) - 0 | = \frac{1}{2\sqrt{n}}$$

and $\lim_{n \to \infty} \| f_n - f \| = 0$, so $f_n \to 0$ in the norm of $C[0, 1]$, as well as pointwise.

Convergence in the sup norm may be approached from a somewhat different point of view but with an equivalent result. Let us examine the definition of $\lim_{n \to \infty} f_n(x) = f(x)$ which says that for every x the following situation exists: Given $\epsilon > 0$, there exists an N such that for $n > N$, $| f_n(x) - f(x) | < \epsilon$. Here it is emphasized that x is designated first, and it is therefore reasonable to expect that the N found later may depend on both x and ϵ. This is in fact the case in Example 3.2.2 as may be shown by fixing ϵ (say $\epsilon = 0.01$) and then calculating N for various values of x close to 1.

An important situation arises, however, when the procedure just described is not necessary; that is, it is not necessary to know the value of x before determining N, and thus N depends only on ϵ. We then have *uniform convergence*, which is illustrated in Example 3.2.3.

3.2.4   *DEFINITION*   Let $\{f_n\}$ be a sequence of functions with respective domains D_n; let f be the pointwise limit function with domain $D = \cap \, D_n$. Then, f_n *converges uniformly* to f for $x \in D$ if, given $\epsilon > 0$, there exists an N such that if $n > N$, $| f_n(x) - f(x) | < \epsilon$.

Graphically this means that for $n > N$, the graphs of $y = f_n(x)$ $(n = 1, 2, \ldots)$ must lie between the graphs of $y = f(x) - \epsilon$ and $y = f(x) + \epsilon$ as shown in Fig. 3. This can obviously happen if and only if $\sup_{D \in x} | f_n(x) - f(x) | < \epsilon$ for $n > N$. Thus *uniform convergence is equivalent to convergence in the sup norm.*

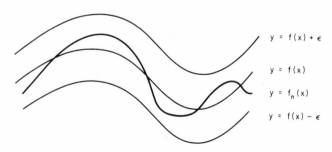

$y = f(x) + \epsilon$

$y = f(x)$

$y = f_n(x)$

$y = f(x) - \epsilon$

Figure 3

PROBLEMS

In the following problems the student will find it instructive to check the analytical results obtained here with the graphs in the Appendix.

In Problems 1–9,

(a) verify that f_n is an element of the space indicated for $n = 1$, $2, \ldots$;
(b) find the pointwise limit f;
(c) determine whether f is an element of the space; and
(d) find $\lim_{n \to \infty} \| f_n - f \|$.

1. $f_n(x) = \dfrac{nx}{1 + n^2 x^2} \in C[0, 1]$.

2. $f_n(x) = n^2 x (1 - x)^n \in C[0, 1]$.

3. $f_n(x) = nx(1 - x)^n \in C[0, 1]$.

4. $f_n(x) = n^2 x (1 - x)^n \in B[0, 1]$.

5. $f_n(x) = xe^{-nx} \in C[0, 1]$.

6. $f_n(x) = \dfrac{1}{n} e^{-nx} \in C^{(1)}[0, 1]$.

7. $f_n(x) = \dfrac{x}{1 + nx^2} \in C^{(1)}[0, 1]$.

8. $f_n(x) = nxe^{-nx} \in C[1, \infty)$.

9. $f_n(x) = nxe^{-nx} \in C[0, \infty)$.

10. In which of the Problems 1–5 is convergence uniform?

11. Let $f_n(x) = x^n$ for $0 \le x \le 1$, and $\lim_{n \to \infty} f_n(x) = f(x)$. The function f was found in Example 3.2.2. Fix $\epsilon = 0.01$. Calculate N in the definition of $\lim_{n \to \infty} f_n(x) = f(x)$ for $x = 0.9$, 0.99, and 0.999.

12. Do Problem 11, where $f_n(x) = nx(1 - x)^n$ as in Problem 3, with $\epsilon = 0.01$ and $x = 0.1$, 0.01, and 0.001. (A computer is helpful.)

13. Recall that uniform convergence is the same as convergence in the sup norm. However, in Problems 6 and 7, the norm convergence required is more than just the sup norm. Thus in

these problems we could have uniform convergence and still not have the norm convergence required. Does this in fact happen in either case?

14. The definition of a Cauchy sequence of numbers was given in Definition 3.1.6. Define a Cauchy *pointwise* sequence of functions and a Cauchy *uniform* sequence of functions, in terms of ϵ, m, n, and N.

15. In an indirect proof, it is necessary to assume the negation of the proposition to be proved. State the negation of uniform convergence.

16. Recall that a function f is bounded if $\{f(x)\}$ is bounded and thus $f_n(x)$ may be bounded for each n. Devise a definition for *uniform boundedness* of a sequence. (Note: Here "n" plays the same part as "x" does in Definition 3.2.4.)

17. In Problems 3 and 4, determine whether the sequences are *uniformly* bounded.

3.3 completeness of sequence spaces

The sequence spaces considered in Chapter II provided a neat package for the study of absolutely convergent series; however, their importance does not stop there. Specific spaces (l_2, for example, in the role of a representation of Hilbert space) have many important applications. Further, these spaces of sequences behave very much like spaces of integrable functions. Thus, a little familiarity gained by considering the completeness of these sequence spaces is well worth the effort here.

The solution to Problem 7, Section 2.2, forms a pattern for the general proof of completeness of (c) and (c_0). We give here a proof of the completeness of l_p and leave the other proofs as problems.

3.3.1 THEOREM   The space l_p, $1 \leq p < \infty$, is complete.

Proof   Let $\{x_n\}$ be a Cauchy sequence in l_p, that is, $\| x_n - x_m \| \to 0$ as m, $n \to \infty$, and

$$x_1 = (\alpha_1^{(1)}, \alpha_2^{(1)}, \ldots, \alpha_k^{(1)}, \ldots),$$
$$x_2 = (\alpha_1^{(2)}, \alpha_2^{(2)}, \ldots, \alpha_k^{(2)}, \ldots),$$
$$x_3 = (\alpha_1^{(3)}, \alpha_2^{(3)}, \ldots, \alpha_k^{(3)}, \ldots),$$
$$\vdots$$
$$x_n = (\alpha_1^{(n)}, \alpha_2^{(n)}, \ldots, \alpha_k^{(n)}, \ldots),$$
$$\vdots$$

Now, fix k.

$$|\alpha_k^{(n)} - \alpha_k^{(m)}| \leq \left(\sum_{i=1}^{\infty} |\alpha_i^{(n)} - \alpha_i^{(m)}|^p\right)^{1/p}$$

$$= \|x_n - x_m\|.$$

So, $\{\alpha_k^{(n)}\}$ is a Cauchy sequence of real numbers.

Now, let $\alpha_k = \lim_{n \to \infty} \alpha_k^{(n)}$. As in the proof of Theorem 3.1.8(a), $\{\|x_n\|\}$ is bounded, say, $\|x_n\| \leq M$. For any k,

$$\left(\sum_{i=1}^{k} |\alpha_i^{(n)}|^p\right)^{1/p} \leq \|x_n\| \leq M.$$

Now let $n \to \infty$ and we have

$$\left(\sum_{i=1}^{k} |\alpha_i|^p\right)^{1/p} \leq M.$$

However, k is arbitrary, so the infinite sum must be less than or equal to M, and thus $\{\alpha_k\} \in l_p$.

Now, let $x = \{\alpha_k\}$. We shall show that $\|x_n - x\| \to 0$. Fix $\epsilon > 0$. There exists an N such that $\|x_n - x_m\| < \epsilon$ if $m, n > N$. Now, for every k,

$$\left(\sum_{i=1}^{k} |\alpha_i^{(n)} - \alpha_i^{(m)}|^p\right)^{1/p} \leq \|x_n - x_m\| < \epsilon.$$

Let $m \to \infty$.

$$\left(\sum_{i=1}^{k} |\alpha_i^{(n)} - \alpha_i|^p\right)^{1/p} \leq \epsilon.$$

This is true for every k, so

$$\left(\sum_{i=1}^{\infty} |\alpha_i^{(n)} - \alpha_i|^p\right)^{1/p} \leq \epsilon.$$

That is, $\|x_n - x\| \leq \epsilon$. ∎

PROBLEMS

1. Prove the following linear spaces are complete:

   (a) the space (c_0),
   (b) the space (c),
   (c) the space $l_p{}^n$,
   (d) the space l_∞,
   (e) the space $l_\infty{}^n$.

   (*Hint*: for (a) and (b) see Problems 7 and 8 in Section 2.2.)

2. Give a separate proof [not just a corollary of Problem 1(c)] that E^2 is complete.

3. Let X be a metric space with a distance function $d(x, y)$ as in Problem 13, Section 1.2. Using Cauchy sequences (Problem 7, Section 3.1), define a *complete metric space*.

4. Let x be a sequence $\{\alpha_n\}$ of real numbers. Define

$$// \, x \, // = \sum_{n=1}^{\infty} \frac{1}{2^n} \frac{|\,\alpha_n\,|}{1 + |\,\alpha_n\,|}.$$

   Why is $// \, x \, //$ not a norm?

5. In Problem 4, define $d(x, y) = // \, x - y \, //$. Show that X is then a complete metric space but not a Banach space. Such a space is called a *Frechet space*.

6. As a consequence of Problem 1(c) prove that if $\{\alpha_{n,k}\}$ is a double sequence of positive terms,

$$\lim_{n \to \infty} (\lim_{k \to \infty} \alpha_{n,k}) = \lim_{k \to \infty} (\lim_{n \to \infty} \alpha_{n,k}),$$

   where $+ \infty$ may be allowed as a limit.

IV

CONTINUOUS
FUNCTIONS

4.1 completeness of C[a, b]—uniform continuity

We recall that f is continuous at c if $\lim_{x \to c} f(x) = f(c)$, that is, if for every $\epsilon > 0$ there exists a $\delta > 0$ such that $|f(x) - f(c)| < \epsilon$ when $|x - c| < \delta$. Note that this must hold even when $x = c$. For continuity at end points of an interval we must obviously appeal to the concept of a one-sided limit.

4.1.1 *DEFINITION* The function f is continuous at c on the right (respectively, left) if for every $\epsilon > 0$ there exists a $\delta > 0$ such that $|f(x) - f(c)| < \epsilon$ when $x - c < \delta$ (respectively, $c - x < \delta$). We write $\lim_{x \to c+} f(x) = f(c)$ [respectively, $\lim_{x \to c-} f(x) = f(c)$].

The function f is continuous in $[a, b]$ if it is continuous at every point in $[a, b]$ with continuity at end points defined in Definition 4.1.1.

We have anticipated the fact that the space of continuous functions is complete and now prove it as a theorem.

4.1.2   *THEOREM*   The space $C[a, b]$ is complete with the norm $\| f \| = \sup_{a \leq x \leq b} | f(x) |$.

Proof   Let

$$\lim_{m, n \to \infty} \| f_m - f_n \| = 0, \qquad f_m, f_n \in C[a, b].$$

Since $\| g \| = \sup_x | g(x) |$, the Cauchy sequence $\{ f_n \}$ gives rise to a Cauchy sequence of real numbers $\{ f_n(x) \}$ for each x, which converges. Let $f(x)$ be that pointwise limit. We claim that $\lim_{n \to \infty} \| f_n - f \| = 0$. For, suppose not. Then there is an $\epsilon > 0$ such that the following holds: For every N there is a $k > N$ and an x_0 such that $| f_k(x_0) - f(x_0) | \geq \epsilon$. Fix $\epsilon > 0$. Now, there exists an N' such that $\| f_m - f_n \| < \epsilon/4$ for $m, n > N'$. Let this be the N referred to above and let it be fixed; this determines k and x_0. (See Fig. 4.) Now, f is the pointwise limit of $\{ f_n \}$, so for this x_0 there exists N'' such that $| f_n(x_0) - f(x_0) | < \epsilon/4$ if $n > N''$. Since f_k is con-

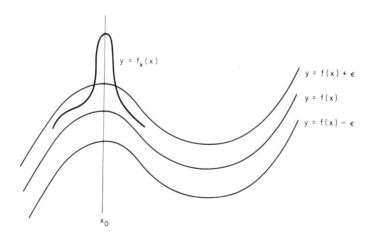

y = f_k(x)

y = f(x) + ε

y = f(x)

y = f(x) − ε

x_0

Figure 4

tinuous, there exists $\delta_1 > 0$ such that if $|x - x_0| < \delta_1$, $|f_k(x) - f_k(x_0)| < \epsilon/4$. Now, let $m > N = \max(N', N'')$. Then, since f_m is continuous, there exists $\delta_2 > 0$ such that if $|x - x_0| < \delta_2, |f_m(x) - f_m(x_0)| < \epsilon/4$. Take $|x - x_0| < \delta = \min(\delta_1, \delta_2)$. Then,

$$|f_k(x_0) - f(x_0)| < |f_k(x_0) - f_k(x)| + |f_k(x) - f_m(x)|$$

$$+ |f_m(x) - f_m(x_0)| + |f_m(x_0) - f(x_0)|$$

$$< \frac{\epsilon}{4} + \frac{\epsilon}{4} + \frac{\epsilon}{4} + \frac{\epsilon}{4} = \epsilon.$$

This gives a contradiction. Thus, $\lim_{n \to \infty} \|f_n - f\| = 0$. It remains to be shown that $f \in C[a, b]$. Fix $\epsilon > 0$ and let $a \le x_0 \le b$. Then there exists an M such that for $n > M$, $|f_n(x_0) - f(x_0)| < \epsilon/3$. By the continuity of f_{M+1} there exists a $\delta > 0$ such that $|f_{M+1}(x_0) - f_{M+1}(x)| < \epsilon/3$ when $|x - x_0| < \delta$. Then, for $|x - x_0| < \delta$,

$$|f(x) - f(x_0)| \le |f(x) - f_{M+1}(x)| + |f_{M+1}(x) - f_{M+1}(x_0)|$$

$$+ |f_{M+1}(x_0) - f(x_0)|$$

$$< \frac{\epsilon}{3} + \frac{\epsilon}{3} + \frac{\epsilon}{3} = \epsilon.$$

Thus f is continuous at x_0, and since x_0 was any point in $[a, b], f \in C[a, b]$. ∎

It can be seen that continuity at a point and convergence at a point are both local properties, and the definition of continuity in an interval does nothing but extend this local property to all the points in the interval. The same can be said for pointwise convergence of a sequence in an interval. However, the concept of norm convergence, or uniform convergence, introduces the idea of convergence on a *set* of points rather than convergence at lots of points. Similarly, the idea of continuity can be extended to a property of a function on a set rather than at several points.

4.1.3 *DEFINITION* The function f is *uniformly continuous* on a set S if given $\epsilon > 0$, there is a $\delta > 0$ such that $|f(x) - f(c)| < \epsilon$ whenever $|x - c| < \delta$ for all x and $c \in S$.

Note the consistency in the use of the word "uniformly." Here we do not designate the point c before finding δ; and in uniform convergence we do not designate the point x before finding the N such that if $n > N$, $|f_n(x) - f(x)| < \epsilon$. Thus we refer to the "uniform δ" in the first case and the "uniform N" in the latter. Of course, there would be no trouble in finding the uniform δ or the uniform N if we were dealing with only a finite number of points c or x, for we would simply take the smallest of all the δ's corresponding to the c's, and the largest of the N's corresponding to the x's.

The following theorem shows how, in one case at least, we can replace an infinite number of intervals like $(c - \delta, c + \delta)$ by a finite number.

4.1.4 *THEOREM* (*Heine–Borel theorem*) Let $[a, b]$ be a closed bounded interval. Let $\{I_\alpha\}$ be an infinite collection of open intervals that cover $[a, b]$; that is, if $x \in [a, b]$, $x \in I_\alpha$ for some α. Then a finite number of I_α, say, $I_{\alpha_1}, I_{\alpha_2}, \ldots, I_{\alpha_k}$ cover $[a, b]$. (Note that $\{I_{\alpha_k}\}$ is not a new, but finite, collection; each I_{α_k} is some I_α.)

Proof Suppose there is not a finite subset of $\{I_\alpha\}$ that will cover $[a, b]$. Then it must require an infinite number of intervals of the collection $\{I_\alpha\}$ to cover either $[a, (a + b)/2]$, or $[(a + b)/2, b]$. Choose the one that requires an infinite number (if both require it, choose either), and repeat the construction, this time finding an interval of length $(b - a)/4$ requiring an infinite number of the $\{I_\alpha\}$ intervals. Continuing, we will have intervals of lengths $(b - a)/2^n$, each included in the preceding interval. The right-hand end points form a Cauchy sequence of real numbers which has a limit, say, c, as n approaches infinity. Now, $a \leq c \leq b$, and c is in some interval I_c of the collection $\{I_\alpha\}$. Let the length of this interval be $\delta = \delta(I_c)$. Now, take n large enough so that $(b - a)/2^n < \delta$. By definition the interval of length $(b - a)/2^n$ requires an infinite number of intervals of $\{I_\alpha\}$ to cover it, but we have found a single one to cover it and thus have a contradiction, which proves the theorem. ∎

4.1.5   THEOREM   If f is continuous on a *closed* interval $[a, b]$, f is uniformly continuous on $[a, b]$.

Proof   Fix $\epsilon > 0$. Since f is continuous at every point c, $a \le c \le b$, there is a $\delta(c)$ such that $|f(x) - f(c)| < \epsilon/2$ if $|x - c| < \delta(c)$. Now, with each point c we associate an interval I_c whose midpoint is c and whose *total length* (not half-length) is $\delta(c)$. These intervals $\{I_c\}$ cover $[a, b]$. (For the points a and b we may either consider intervals extending beyond $[a, b]$ or take the intersection of these intervals with $[a, b]$ for new intervals.) By the Heine–Borel theorem a finite number of these intervals, say I_{c_1}, I_{c_2}, ..., I_{c_k} cover $[a, b]$. Let $\delta(c_k)$ be the length of I_{c_k} and let $\delta = \frac{1}{2} \min[\delta(c_1),$ $\delta(c_2),$ ..., $\delta(c_k)]$. Then, if $x_0 \in [a, b]$, x_0 lies in some I_{c_k} whose midpoint is c_k, so

$$|x_0 - c_k| < \tfrac{1}{2}\delta(c_k).$$

Now, if x is any point so that $|x - x_0| < \delta$, then

$$|x - c_k| \le |x - x_0| + |x_0 - c_k|$$
$$< \delta + \tfrac{1}{2}\delta(c_k) \le \tfrac{1}{2}\delta(c_k) + \tfrac{1}{2}\delta(c_k) = \delta(c_k).$$

So,
$$|f(x) - f(x_0)| \le |f(x) - f(c_k)| + |f(c_k) - f(x_0)|$$

$$< \frac{\epsilon}{2} + \frac{\epsilon}{2} = \epsilon. \ \blacksquare$$

A theorem closely related to the Heine–Borel theorem is the Bolzano–Weierstrass theorem:

4.1.6   THEOREM   (*Bolzano–Weierstrass theorem*)   Every bounded infinite set of real numbers has a limit point.

Proof   Since the set is bounded, it is contained in some interval $[a, b]$. Construct intervals, as in the proof of the

Heine–Borel theorem, of lengths $(b - a)/2^n$, and call the nth one A_n. Then, $A_n \subset A_{n+1}$, and the left-hand end points of these intervals form a bounded monotone nondecreasing sequence which has a limit, say, α. It is easy to verify that α is a limit point of the original set. ∎

We may suppose that continuity is such a strong condition that it forces boundedness on a function. Such is indeed the case, as the following theorem shows, if the interval is closed. To see the necessity of a closed interval, consider $f(x) = 1/x$ for $0 < x < 1$. Certainly f is continuous in $(0, 1)$ but not bounded.

4.1.7 THEOREM

$$C[a, b] \subset B[a, b].$$

Proof Assume $f(x)$ is continuous but not bounded above. Then, for a given M, there is an x_1 such that $f(x_1) > M + 1$. Similarly, there is an x_2 such that $f(x_2) > f(x_1) + 1$. Continuing, $f(x_{n+1}) > f(x_n) + 1$, and $\{x_n\}$ is an infinite sequence which, by the Bolzano–Weierstrass theorem, has a limit point, say, x_0. Now, f is continuous at x_0, so for $\epsilon = \frac{1}{4}$, there is a $\delta > 0$ such that if $|x - x_0| < \delta$, $|f(x) - f(x_0)| < \frac{1}{4}$. Since x_0 is a limit point of $\{x_n\}$, there are $x_{n'}$ and $x_{n''}$ such that $|x_{n'} - x_0| < \delta$ and $|x_{n''} - x_0| < \delta$. Then,

$$\begin{aligned} |f(x_{n'}) - f(x_{n''})| &\leq |f(x_{n'}) - f(x_0)| \\ &\quad + |f(x_0) - f(x_{n''})| \\ &< \tfrac{1}{4} + \tfrac{1}{4} = \tfrac{1}{2}. \end{aligned}$$

But, by construction, $|f(x_{n'}) - f(x_{n''})| > 1$, so the theorem is proved by contradiction. ∎

PROBLEMS

1. Using the definition of continuity, prove that if f and g are continuous on $[a, b]$, then $\alpha f + \beta g$ is continuous on $[a, b]$. (See Problem 1, Section 1.1.)

2. If $f \in C[a, b]$ and $g \in C[a, b]$, prove $fg \in C[a, b]$.

3. If $f \in C[a, b]$ and $g \in C[a, b]$, prove $f/g \in C[a, b]$ provided $g(t)$ is not zero for any $t \in [a, b]$.

4. If $f \in C[a, b]$ and $f + g \in C[a, b]$, prove $g \in C[a, b]$ or give an example where $g \notin C[a, b]$.

5. If $f \in C[a, b]$ and $fg \in C[a, b]$, prove $g \in C[a, b]$ or give an example where $g \notin C[a, b]$.

6. If $f + g \in C[a, b]$, do we necessarily have $f \in C[a, b]$, $g \in C[a, b]$? Give an example.

7. If f is continuous on $[a, b]$ except at c, and if f may be made continuous there by defining $f(c) = k$, c is called a *removable discontinuity*. Which of the following have removable discontinuities? Define the functions at those points so as to make them continuous.

  (a) $\dfrac{x^3 - 8}{x - 2}$, $0 \le x \le 4$,

  (b) $\dfrac{\sqrt{1 + x} - 1}{x}$, $-1 \le x \le 1$,

  (c) $\dfrac{1}{x}$, $-1 \le x \le 1$,

  (d) $\sqrt{\dfrac{1}{x^2} + \dfrac{2}{x}} - \dfrac{1}{x}$, $-1 \le x \le 1$.

8. Let $f(t)$ be defined for $a \le t \le b$, and let $\lim_{t \to c} f(t) = L$. Let $g(s)$ be defined on some interval containing L, and let $\lim_{s \to L} g(s) = K$. Assume $g(s)$ is continuous for $s = L$. Prove $\lim_{t \to c} g(f(t)) = K$.

9. Let

$$f(x) = \begin{cases} x \sin \dfrac{1}{x} & \text{if } 0 < x < \dfrac{2}{\pi}, \\ 0 & \text{if } x = 0. \end{cases}$$

Prove from the definition that f is continuous on the right at zero.

10. Give an example to show that in Problem 8, g must be continuous at L in order to have the conclusion valid. (A function similar to that in Problem 9 may be useful.)

11. Determine whether the following are uniformly continuous on the interval stated. Give reasons.

(a) $f(x) = x^2$, $0 \le x < \infty$,

(b) $f(x) = x \sin \dfrac{1}{x}$, $0 < x \le \dfrac{2}{\pi}$,

(c) $f(x) = \dfrac{1}{x}$, $0 < x \le 1$,

(d) $f(x) = \begin{cases} \dfrac{1}{x-1}, & 0 \le x \le 2, \quad x \ne 1, \\[2mm] 0, & x = 1, \end{cases}$

(e) $f(x) = \begin{cases} e^{-1/x^2}, & -1 < x < 1, \quad x \ne 0, \\[2mm] 0, & x = 0, \end{cases}$

(f) $f(x) = \begin{cases} e^{-1/x^2}, & -\infty < x < \infty, \quad x \ne 0, \\[2mm] 0, & x = 0. \end{cases}$

12. (a) Show by the definition that $f(x) = 1/x$ is uniformly continuous on $[\frac{1}{2}, 1]$.

(b) Adapt the proof in (a) to show that $f(x)$ is uniformly continuous on $[\eta, 1]$, where $0 < \eta < 1$.

13. (a) Do Problem 12(a) for $f(x) = x^2$ on $[0, 100]$.

(b) Do Problem 12(b) for $f(x) = x^2$ on $[0, M]$.

14. Prove or disprove that the condition in Theorem 4.1.7 that the interval be closed may be replaced by the condition that f be *uniformly* continuous on the interval (whether open or closed).

15. Let

$$f(x) = \begin{cases} x & \text{if } x \text{ is rational,} \quad 0 \leq x \leq 1, \\ 1 - x & \text{if } x \text{ is irrational,} \quad 0 \leq x \leq 1. \end{cases}$$

For what values of x is f continuous? Give a proof.

16. Let

$$f(x) = \begin{cases} 0 \text{ if } x \text{ is irrational,} \\ \dfrac{1}{q} \text{ if } x = \dfrac{p}{q} \text{ and } p \text{ and } q \text{ have no common factor.} \end{cases}$$

For what values of x is f continuous? Give a proof.

4.2 properties of continuous functions

Many properties of continuous functions are so basic that they are sometimes implicitly assumed to be true for all functions. For example, in estimating roots of an equation $f(x) = 0$ we may find that $f(a)$ is negative and $f(b)$ is positive and so assume that there is a root of the equation between a and b. This is true if f is a continuous function but not necessarily true otherwise. [Try $f(x) = 1/x$, where $f(1) = 1, f(-1) = -1$.]

It should be clear from some of the previous discussions that these "obvious" conclusions usually depend in some way on the completeness of the real numbers (both the domain and the range of the function) and on the preservation of "closeness," which is a characteristic of continuous functions. We give here a few of these familiar properties as theorems and additional ones in the problems.

4.2.1 *THEOREM* If $f \in C[a,b]$, and if $M = \sup_{a \leq x \leq b} f(x)$, there is a value x_0, $a \leq x_0 \leq b$, such that $f(x_0) = M$.

Proof Suppose there is no value x_0 such that $a \leq x_0 \leq b$ and $f(x_0) = M$. Then $M - f(x) > 0$ for all x in $[a, b]$ so

$g(x) = 1/[M - f(x)]$ is continuous for $x \in [a, b]$. Therefore, by Theorem 4.1.7, g is bounded. But this is a contradic-
tion of the fact that $M = \sup_{a \le x \le b} f(x)$; for if $1/[M - f(x)] \le A$, $f(x) \le M - (1/A)$. This contradiction proves the
theorem. ∎

As a result of this theorem we could use $\| f \| = \max | f(x) |$ in-
stead of $\sup | f(x) |$ for the norm in $C[a, b]$.

4.2.2 THEOREM   Let f be continuous at c and let $f(c) > 0$.
Then there is an interval $(c - \delta, c + \delta)$ such that if $x \in (c - \delta, c + \delta), f(x) > 0$.

Proof   Let $\epsilon = \frac{1}{2}f(c)$. By the continuity of f there is a
δ such that if $| x - c | < \delta$, $| f(x) - f(c) | < \frac{1}{2}f(c)$. Thus
$\frac{1}{2}f(c) < f(x) < \frac{3}{2}f(c)$ so $f(x) > 0$. ∎

4.2.3 THEOREM   If f is continuous on any interval containing
a and b, and if $f(a) < 0, f(b) > 0$, there is a value $c, a < c < b$,
such that $f(c) = 0$.

Proof   Let $S = \{x \mid a \le x \le b, \ f(x) \le 0\}$. Then S is
nonempty and bounded. Let $c = \sup S$. Then $f(c) = 0$. For,
if $f(c) > 0$, there is an interval around c, say, $(c - \delta, c + \delta)$,
such that if $x \in (c - \delta, c + \delta)$, $f(x) > 0$. (The interval is
modified in the obvious way if $c = b$.) Let $c - \delta < \xi < c$, so
$f(\xi) > 0$. This is a contradiction of the definition of $c = \sup S$.
Similarly, $f(c)$ cannot be less than 0. Thus, $f(c) = 0$, and the
theorem is proved. ∎

4.2.4 THEOREM   Let f be continuous on any interval containing
a and b with $a < b$ and $f(a) < f(b)$. If $f(a) < K < f(b)$,
there is a value $c, a < c < b$, such that $f(c) = K$.

Proof   Let $g(x) = f(x) - K$. Then use the previous
theorem. ∎

Monotone functions are defined in a way similar to that for sequences, and lim sup and lim inf have meanings compatible to those
for sequences as will be shown in the problems.

4.2.5 *DEFINITION*   If for all x_1 and x_2 in an interval with $x_1 < x_2$, we have $f(x_1) < f(x_2)$, f is *monotone increasing* in the interval; if $f(x_1) \leq f(x_2)$, it is *monotone nondecreasing;* if $f(x_1) > f(x_2)$, it is *monotone decreasing;* if $f(x_1) \geq f(x_2)$, it is *monotone nonincreasing.*

4.2.6 *DEFINITION*   Let

$$u(\delta) = \sup_{0<|x-c|<\delta} f(x) \quad \text{and} \quad l(\delta) = \inf_{0<|x-c|<\delta} f(x).$$

Then

$$\limsup_{x \to c} f(x) = \lim_{\delta \to 0+} u(\delta) \quad \text{and} \quad \liminf_{x \to c} f(x) = \lim_{\delta \to 0+} l(\delta).$$

PROBLEMS

1. Let

$$f_n(x) = \frac{x^n}{1 + x^n}, \qquad 0 \leq x \leq 1.$$

Note that $\sup_{0\leq x\leq 1} f_n(x) = \frac{1}{2}$ and $\inf_{0\leq x\leq 1} f_n(x) = 0$. (See Appendix for graphs.) Show $f_n \in C[0, 1]$. Find the value c in Theorem 4.2.4 so that $f_n(c) = \frac{1}{4}$. What happens when $n \to \infty$? Explain. (*Hint:* consider whether $\{f_n\}$ converges in the norm of $C[0, 1]$.)

2. Define $[x]$ to be the greatest integer less than or equal to x. Discuss left continuity and right continuity for $f(x) = [x]$.

3. Let

$$f_n(x) = x^{1/(2n-1)} \quad \text{and} \quad f(x) = \lim_{n \to \infty} f_n(x), \quad -1 \leq x \leq 1.$$

Discuss left and right continuity of f at the origin. Does $f_n \to f$ in the sup norm?

4. Show that the following definitions agree with those in Definition 4.2.6.

$$\limsup_{x \to c} f(x) = \inf_{\delta>0} [\sup_{0<|x-c|<\delta} f(x)],$$

$$\liminf_{x \to c} f(x) = \sup_{\delta > 0} [\inf_{0 < |x - c| < \delta} f(x)].$$

5. State definitions for lim sup and lim inf of $f(x)$ as $x \to \infty$. If $f(n) = a_n$, and $\{a_n\}$ is considered as a sequence, show that these definitions agree with Definition 2.1.5.

6. Find lim inf and lim sup for the following:

(a) $\dfrac{1}{x+1} \sin \dfrac{1}{x}$ as $x \to 0$.

(b) $\dfrac{x}{x+1} \sin x$ as $x \to \infty$.

(c) $\cos \dfrac{1}{x}$ as $x \to 0$.

7. If $\limsup_{x \to c} f(x) = \liminf_{x \to c} f(x) = f(c)$, prove that f is continuous at c.

8. If n is a positive integer and $a > 0$, prove there is exactly one positive b such that $b^n = a$.

9. Let $f \in C[0, 1]$, $0 \le f(x) \le 1$. Prove that there is at least one point c, $0 \le c \le 1$, such that $f(c) = c$.

10. Prove Theorem 4.2.1 as follows: construct a sequence $\{x_n\}$ such that $f(x_n) > M - (1/n)$. Let $x_0 = \limsup_{n \to \infty} x_n$.

In Problems 11–14 show that $f_n \in C[0, 1]$, $n = 1, 2, \ldots$, find $f(x) = \lim_{n \to \infty} f_n(x)$ and determine whether the convergence is uniform, that is, whether $\| f_n - f \| \to 0$ as $n \to \infty$. (See Appendix for graphs.)

11. $f_n(x) = x^n + x^2$.

12. $f_n(x) = x(e^{-nx} + x)$.

13. $f_n(x) = nxe^{-nx} + \dfrac{e^{x-1}}{n}$.

14. $f_n(x) = \dfrac{2x + nx^3}{1 + nx^2}$.

4.3 some topological concepts

We have thus far discussed sets of real numbers and sets of points in various spaces with only as much reference to topological concepts as was necessary to make the meaning clear. This has allowed us to formulate important results without an undue amount of detail but has sacrificed some generality. For example, in the proof of the Heine–Borel theorem it is essential that the point set which we called $[a, b]$ be closed and bounded but not that it be an interval. If it is bounded, it is contained in an interval $[a, b]$, and the construction proceeds as before.

The sets of real numbers that we have used have usually been open or closed intervals. Starting with these basic concepts, however, more general sets can be defined; and results obtained for numbers on the real line can often be extended to the plane and other spaces with little or no change in terminology.

We start with the notion of an open interval. However, since that terminology ties us to the real line, the word "neighborhood" is used with the understanding that it can be extended to a circle in two dimensions, a sphere in three dimensions, and even more general concepts. For example, if x is a point in a normed linear space, the set $\{y \mid \|x - y\| < \epsilon\}$ is called an *(epsilon) neighborhood of x*. (See Problem 10 for another terminology.)

4.3.1 *DEFINITION* A set S is *open* if for every $x_0 \in S$ there is a neighborhood containing x_0 and entirely contained in S.

When the point x_0 can be put in a neighborhood as in the previous definition, it is called an *interior point* of S. Thus, an open set consists entirely of interior points.

4.3.2 *DEFINITION* The *complement* of a set S consists of all elements not in S; we write CS.

It should be clear that in order for CS to be well defined we need to know what the whole "universe" is in any given problem. This is usually clear, but it often pays to ask "complement with respect to what"?

4.3.3 DEFINITION   A set T is *closed* if and only if CT is open.

The definition of a limit point can be taken directly from Definition 2.1.6 by simply changing "interval" to "neighborhood." Thus, for any set S, x_0 is a limit point if every neighborhood of x_0 contains points of S other than x_0 itself.

4.3.4 THEOREM   A closed set contains all its limit points.

Proof   Let x_0 be a limit point of a closed set T. If $x_0 \in CT$, there must be a neighborhood of x_0 entirely in CT. But this is impossible if x_0 is a limit point of T.   ▮

4.3.5 REMARK   We should note here that the above procedure could have been reversed in that we could have defined a closed set as in Theorem 4.3.4 and an open set to be the complement of a closed set; then Definition 4.3.1 would follow as a theorem. Also note that just as in the case of intervals, it cannot be said that every set is either open or closed.

4.3.6 THEOREM   For the real line the following hold:

(a) The whole space and the empty set are both open and closed,
(b) An open interval is an open set,
(c) A closed interval is a closed set,
(d) The union of any number of open sets is open,
(e) The intersection of a finite number of open sets is open,
(f) The intersection of any number of closed sets is closed,
(g) The union of a finite number of closed sets is closed.

Proof   The proof is left as Problems 1 and 2.   ▮

One general topological concept that carries over directly from a property of real numbers is that of compactness:

4.3.7 DEFINITION   A space (or set) S is *compact* if every collection of open sets that covers S has a finite subcollection that also covers S.

A trivial result, then, is that every closed bounded set of real numbers is compact, because that is just what the Heine–Borel theorem says. In fact, compactness is sometimes referred to as the "Heine–Borel property." This is a fine example of the abstraction of an idea from rather basic concepts, since the Heine–Borel theorem first came about, as we have used it here, to clarify properties of continuous functions.

PROBLEMS

1. Prove parts (a)–(g) of Theorem 4.3.6.

2. Give examples to show that parts (e) and (g) of Theorem 4.3.6 are not true if we replace "finite" by "infinite."

3. If A is a set of points, define A' to be the set of limit points of A. Give examples of sets A satisfying the following:
   (a) $A = A'$
   (b) $A' = \phi$ (the empty set),
   (c) $A \cap A' = \phi$.

4. The *closure* of A is $\bar{A} = A \cup A'$. Prove $\bar{A}$ is closed.

5. A set A is *dense* in X if $\bar{A} = X$. Give three examples each of dense subsets of E^1 and E^2.

6. A space is *separable* if it contains a countable dense subset. Show E^1 and E^2 are separable.

7. The *boundary* of a set A in a space X is the set of all points which are not interior to A nor CA.
   (a) Prove that a set A is closed if and only if A contains its boundary.
   (b) Prove that a set A is open if and only if CA contains the boundary of A.
   (c) What is the boundary of $(0, 1)$, of $(0, 1]$, and of $[0, 1]$?

8. Formulate the definition of a *compact metric space*.

9. Are the following sets of points open or closed or both or neither in E^2?
   (a) $S = \{(x, y) \mid y^2 \leq x^4 - x^2\}$,

(b) $T = \{(x, y) \mid \ln(x^2 + y^2 - 24) > 0\}$,

(c) $U = \{(x, y) \mid \dfrac{y}{x^2 - 4y} \geq 0\}$.

10. Let X be a normed linear space and $x_0 \in X$. A *closed ball* is the set of elements $\{x\}$ such that $\| x - x_0 \| \leq \epsilon$. An *open ball* is the set of elements $\{x\}$ such that $\| x - x_0 \| < \epsilon$ for any ϵ. Using these definitions, define open set, closed set, and limit point for a set in X.

11. State the Heine–Borel theorem for a normed linear space using the terms defined in Problem 10. Is it true?

12. Discuss the following statement: "let $C[a, b]$ be considered a subset in $B[a, b]$. Then $C[a, b]$ is closed in $B[a, b]$."

13. Prove the following: let G be an open set of real numbers and f a real-valued function of real numbers. Then f is continuous if and only $f^{-1}(G)$ is open.

4.4 functions of two variables

If $u = f(x, y)$, problems of continuity and, later, differentiability offer no difficulty when viewed as two iterated operations; that is, we know the meaning of

$$\lim_{y \to y_0} \left(\lim_{x \to x_0} f(x, y) \right) \quad \text{and} \quad \lim_{x \to x_0} \left(\lim_{y \to y_0} f(x, y) \right).$$

Unfortunately, they are not always the same, as the example

$$u = \frac{x^2 - y^2}{x^2 + y^2}$$

shows. What, then, do we mean by

$$\lim_{(x,y) \to (x_0, y_0)} f(x, y) \; ?$$

Let us rephrase the problem in terms of linear spaces. To this end, let $x = (\alpha, \beta)$ be a point in $E^2 (= l_2^2)$, and let $u = f(x)$ be a real-valued function (that is, $u \in E^1$). The following definitions should be obvious ones if we simply replace $|x|$ (when $x \in E^1$) by $\|x\|$ (when $x \in E^2$):

4.4.1 *DEFINITION* Let $x = (\alpha,\ \beta)$, $x_0 = (\alpha_0,\ \beta_0)$, and $u = f(x)$. Then, (a) $\lim_{x \to x_0} f(x) = L$ if for every $\epsilon > 0$ there exists a $\delta > 0$ such that $|f(x) - L| < \epsilon$ when $0 < \|x - x_0\| < \delta$, that is, when $\{(\alpha - \alpha_0)^2 + (\beta - \beta_0)^2\}^{1/2} < \delta$. (b) f is continuous at x_0 if $\lim_{x \to x_0} f(x) = f(x_0)$ in the above sense.

We have observed that many different linear spaces have the same points but different norms, for example, $l_2{}^2$ and $l_\infty{}^2$. The norm in $l_\infty{}^2$ is especially easy to use, and it is *equivalent* to the norm in $l_2{}^2$ in the sense that convergence in each norm implies convergence in the other. This is easy to phrase in terms of neighborhoods:

4.4.2 *DEFINITION* Two norms, $\|\cdot\|_1$ and $\|\cdot\|_2$, are *equivalent* if every neighborhood of the origin in each norm contains a neighborhood of the origin in the other norm.

4.4.3 *THEOREM* The norms in l_2^2 and l_∞^2 are equivalent.

Proof Let the norms be denoted by $\|\cdot\|_2$ and $\|\cdot\|_\infty$, and let $x = (\alpha, \beta)$. An ϵ-neighborhood of the origin in $l_2{}^2$ is a circle with radius ϵ, and an ϵ-neighborhood of the origin in $l_\infty{}^2$ is a square with each side ϵ as shown in Fig. 5. It is easy to see

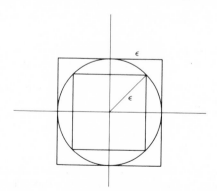

Figure 5

that $(\alpha^2 + \beta^2)^{1/2} < \epsilon$ implies $\sup(\,|\,\alpha\,|,\,|\,\beta\,|\,) < \epsilon$, and that $\sup(\,|\,\alpha\,|,\,|\,\beta\,|\,) < \epsilon/\sqrt{2}$ implies $(\alpha^2 + \beta^2)^{1/2} < \epsilon$. ∎

Returning now to the familiar notation of (x, y) for a point in the plane, we see that $(x, y) \rightarrow (x_0, y_0)$ can mean either $\{(x - x_0)^2 + (y - y_0)^2\}^{1/2} \rightarrow 0$ or $\sup(\,|\,x - x_0\,|,\,|\,y - y_0\,|\,) \rightarrow 0$.

It is easier to prove that $\lim_{(x,y) \rightarrow (x_0,y_0)} f(x, y)$ does not exist than to prove that it does. In the former case we need only find two iterated limits that differ while in the latter case we must show $|f(x, y) - L|$ for all (x, y) in the domain of the function such that $\{(x - x_0)^2 + (y - y_0)^2\}^{1/2} < \delta$ or $\sup(\,|\,x - x_0\,|,\,|\,y - y_0\,|\,) < \delta.$[†]

4.4.4 EXAMPLE Let

$$f(x, y) = \frac{3x^3 - 2y^3}{x^2 + y^2}.$$

Show

$$\lim_{(x,y) \rightarrow (0,0)} f(x, y) = 0.$$

Let $\sup(\,|\,x\,|,\,|\,y\,|\,) = \delta$. Then

$$\left|\frac{3x^3 - 2y^3}{x^2 + y^2}\right| \leq \frac{3\,|\,x\,|^3 + 2\,|\,y\,|^3}{x^2 + y^2} \leq \frac{3\delta x^2 + 2\delta y^2}{x^2 + y^2} \leq 3\delta.$$

Therefore, if $\delta < \frac{1}{3}\epsilon$, $|f(x, y)\,| < \epsilon$.

PROBLEMS

1. Prove that the norms in l_1^2 and l_2^2 are equivalent.

In Problems 2–8, let $(x, y) \rightarrow (0, 0)$. Find $\lim f(x, y)$ if it exists; if it does not, tell why.

2. $f(x, y) = \dfrac{(x + y)^2}{x^2 + y^2}.$ **3.** $f(x, y) = \dfrac{1/x}{(1/x) + (1/y)}.$

[†] The joint limit as $(x, y) \rightarrow (x_0, y_0)$ may exist even though single limits as $x \rightarrow x_0$ or $y \rightarrow y_0$ do not, but it cannot exist if two iterated limits exist and are different.

4. $f(x, y) = \dfrac{x^2 y}{x^4 + y^2}.$

(*Hint*: try
$\lim_{x \to 0} \left[\lim_{y \to x^2} f(x, y) \right]$.)

5. $f(x, y) = \dfrac{x^4 - y^4}{x^2 + y^2}.$

6. $f(x, y) = x \sin \dfrac{1}{y} + y \sin \dfrac{1}{x}.$

7. $f(x, y) = xy \ln(x^2 + y^2).$

8. $f(x, y) = \dfrac{1/x}{(1/y)^2 e^{x/y}}.$

V

DIFFERENTIABLE FUNCTIONS

5.1 preliminary theorems

Let f be defined on $[a, b]$ and let x_0 lie between a and b. If $a < x_0 + h < b$, and if

$$\lim_{h \to 0} \frac{f(x_0 + h) - f(x_0)}{h}$$

exists, this limit is the derivative of f at x_0, written $f'(x_0)$. Alternately,

$$\lim_{x \to x_0} \frac{f(x) - f(x_0)}{x - x_0} = f'(x_0)$$

if this limit exists. Note that we have thus defined a new function f' whose domain is at most the open interval (a, b). One-sided derivatives can be defined at a and b using right- and left-hand limits, respectively, as $h \to 0$ or as $x \to x_0$. We write $f'(a+)$ and $f'(b-)$.

A few fundamental theorems should be reviewed here since there will be occasion to refer to them in the future:

5.1.1 THEOREM If f is differentiable at x_0, f is continuous at x_0.

Proof Write

$$f(x) = \frac{f(x) - f(x_0)}{x - x_0} (x - x_0) + f(x_0).$$

Then

$$\lim_{x \to x_0} f(x) = f'(x_0) \cdot 0 + f(x_0) = f(x_0). \quad \blacksquare$$

5.1.2 THEOREM (*Rolle's theorem*) If f is continuous on $[a, b]$ and differentiable in (a, b), and if $f(a) = f(b) = 0$, there is a value ξ, $a < \xi < b$, such that $f'(\xi) = 0$.

Proof If $f(x) = 0$ for all x between a and b, the theorem is proved. So, let $f(c) \neq 0$ for some c. If $f(c) > 0$, $\sup_{a \le x \le b} f(x) > 0$. Let $\sup_{a \le x \le b} f(x) = M = f(\xi)$ (by Theorem 4.2.1). We claim that $f'(\xi) = 0$. For, $f(x) \le f(\xi)$ for all x in some interval around ξ. If $x < \xi$,

$$\frac{f(x) - f(\xi)}{x - \xi} \ge 0;$$

and if $x > \xi$,

$$\frac{f(x) - f(\xi)}{x - \xi} \le 0.$$

Since

$$\lim_{x \to \xi} \frac{f(x) - f(\xi)}{x - \xi}$$

exists, it must equal zero. If $f(c) < 0$, take $\inf_{a \le x \le b} f(x) = f(\xi)$. An argument similar to that above shows $f'(\xi) = 0$. $\quad \blacksquare$

5.1.3 THEOREM (*Mean value theorem for derivatives*) If f is continuous on $[a, b]$ and differentiable in (a, b), there is a value ξ, $a < \xi < b$, such that

$$f'(\xi) = \frac{f(b) - f(a)}{b - a}.$$

Outline of Proof The proof consists of setting up a function, say, g, in terms of f; g must satisfy the hypotheses of Rolle's theorem; the conclusion of Rolle's theorem, namely, that $g'(\xi) = 0$, would give the desired result for $f'(\xi)$. We are thus looking for a function g whose derivative is zero at a point on the curve where the tangent is parallel to the chord joining $(a, f(a))$ and $(b, f(b))$. Two guesses for g are (1) the length of the chord CD and (2) the area of triangle ABC (see Fig. 6). Both give satisfactory proofs and are suggested in the problems for completing the proof. ∎

5.1.4 *EXAMPLE* Use the mean value theorem to show that $\sin x < x$ if $x > 0$. Let $f(x) = \sin x$ and $x \le 1$. Then, considering the interval $[0, x]$, there is a value ξ, $0 < \xi < x$,

$$\cos \xi = \frac{\sin x - 0}{x - 0}.$$

But, $\cos \xi < 1$ for $0 < \xi < 1$, and the result follows. If $x > 1$, $\sin x \le 1 < x$.

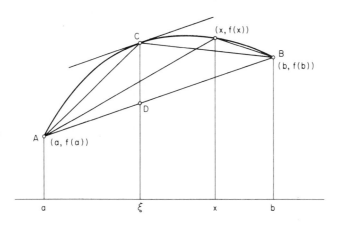

Figure 6

PROBLEMS

1. Complete the proof of Theorem 5.1.3 by letting $g(x)$ be the vertical chord which attains its maximum in the position CD.

2. In Fig. 6 show that the area of the triangle, which attains its maximum in the position ABC, is

$$\frac{1}{2} \begin{vmatrix} a & f(a) & 1 \\ b & f(b) & 1 \\ x & f(x) & 1 \end{vmatrix}$$

and thus we may take

$$g(x) = \begin{vmatrix} a & f(a) & 1 \\ b & f(b) & 1 \\ x & f(x) & 1 \end{vmatrix}.$$

Note that $g(a) = 0 = g(b)$. Then

$$g'(\xi) = \begin{vmatrix} a & f(a) & 1 \\ b & f(b) & 1 \\ 1 & f'(x) & 0 \end{vmatrix} = 0$$

gives the desired result.

3. Let F and G satisfy the hypotheses of the mean value theorem. Let

$$H(x) = \begin{vmatrix} G(a) & F(a) & 1 \\ G(b) & F(b) & 1 \\ G(x) & F(x) & 1 \end{vmatrix}.$$

Show H satisfies the hypotheses of Rolle's theorem. From the conclusion of Rolle's theorem show that there is a value ξ, $a < \xi < b$, such that

$$\frac{F(b) - F(a)}{G(b) - G(a)} = \frac{F'(\xi)}{G'(\xi)}.$$

This is called the generalized mean value theorem.

4. Use the results of Problem 3 to prove the following form of l'Hospital's rule: assume

(1) f and g are defined in some interval including the point c;

(2) they have derivatives f' and g' in this interval;

(3) $g'(x) \neq 0$ in this interval, and

(4) $\lim_{x \to c} f(x) = 0$ and $\lim_{x \to c} g(x) = 0$.

Then, if $\lim_{x \to c} f'(x)/g'(x)$ exists and equals L,

$$\lim_{x \to c} \frac{f(x)}{g(x)} = L.$$

5. Use the mean value theorem to prove that if $f'(x) = 0$ for every x in some interval, then $f(x) = C$ (constant) there.

6. Use the mean value theorem to prove the following inequalities:

(a) $\ln(1 + x) < x$, if $x > 0$.

(b) $\sqrt{1 + x} < 1 + \frac{1}{2}x$, if $x > 0$.

(c) $b \ln b - a \ln a > b - a$, if $1 < a < b$.

7. Prove Lemma 1.3.4 by the mean value theorem.

8. Let f be differentiable for $a < x < b$ and $f'(x) > 0$ for all such x. Prove $f(b) > f(a)$.

9. Let

$$f(x) = \begin{cases} x \cot^{-1} x & \text{if } x \neq 0, \\ 0 & \text{if } x = 0. \end{cases}$$

Find the right- and left-hand derivatives of f at $x = 0$.

10. Let

$$f(x) = \begin{cases} \dfrac{x}{1 + e^{1/x}} & \text{if } x \neq 0, \\ 0 & x = 0. \end{cases}$$

Find the right- and left-hand derivatives of f at 0.

11. Let $x = f^{-1}(y)$ be the inverse function of $y = f(x)$. Show

(a) $\dfrac{d^2x}{dy^2} = -\dfrac{y''}{(y')^3}$,

(b) $\dfrac{d^3x}{dy^3} = \dfrac{3(y'')^2 - y'''y'}{(y')^5}$.

12. Let

$$\overline{D_{x+}} f(x_0) = \limsup_{x \to x_0+} \frac{f(x) - f(x_0)}{x - x_0}$$

with similar definitions for $\overline{D_{x-}}$, $\underline{D_{x+}}$, $\underline{D_{x-}}$. Find all four derivatives at 0 for the function

$$f(x) = \begin{cases} x \sin \dfrac{1}{x} & \text{if } x \neq 0, \\ 0 & \text{if } x = 0. \end{cases}$$

5.2 *completeness of* $C^{(1)}[a, b]$

From elementary calculus we know that the sum, difference, and product of differentiable functions are differentiable; and with a little care to avoid zero in the denominator and even roots of negative numbers, the quotient, powers, and roots of differentiable functions are differentiable. Thus, with the few obvious exceptions, algebraic functions of differentiable functions are differentiable, and the derivatives are given by familiar formulas.

In the case of transcendental functions, however, the situation is somewhat different. The definitions of trigonometric, logarithmic, exponential, hyperbolic, and inverse trigonometric functions all involve, in some way, the idea of a limit of algebraic functions; and, in fact, this is an essential feature of transcendental functions. The study of limits of sequences of functions thus becomes of great importance, and we must look for a type of convergence that will preserve differentiability in the limit function.

The sequence of functions $\{f_n\}$ where $f_n(x) = x^n$, $0 \leq x \leq 1$ shows that pointwise limits of differentiable functions are not always differentiable, and the following example shows that even uniform convergence does not preserve differentiability:

5.2.1 *EXAMPLE*   Let

$$f_n(x) = \frac{x}{1 + nx^2}, \qquad -1 \leq x \leq 1.$$

Then $\lim_{n \to \infty} f_n(x) = f(x) = 0$ for all x in $[-1, +1]$ and

$$\sup_{-1 \leq x \leq +1} |f_n(x) - 0| = \tfrac{1}{2} n^{-1/2} \to 0 \qquad \text{as} \quad n \to \infty.$$

However,

$$f_n'(x) = \frac{1 - nx^2}{(1 + nx^2)^2}$$

and

$$\lim_{n \to \infty} f_n'(x) = \begin{cases} 1 & \text{if} \quad x = 0, \\ 0 & \text{if} \quad x \neq 0. \end{cases}$$

The study of linear spaces should give a clue to the type of convergence that will preserve differentiability in the limit function.

5.2.2 *THEOREM*   Let $C^{(1)}[a, b]$ be the space of functions with continuous first derivatives for $a \leq x \leq b$. Then $C^{(1)}[a, b]$ is complete under the norm

$$\| f \| = \sup_{a \le x \le b} | f(x) | + \sup_{a \le x \le b} | f'(x) |.$$

Proof Let $\{f_n\}$ be a Cauchy sequence in $C^{(1)}[a, b]$. That is, $\| f_m - f_n \| \to 0$ as $m, n, \to \infty$; or

$$\lim_{m,n \to \infty} [\sup | f_m(x) - f_n(x) | + \sup | f_m'(x) - f_n'(x) |] = 0$$

which implies

$$\lim_{m,n \to \infty} \sup | f_m(x) - f_n(x) | = 0$$

and

$$\lim_{m,n \to \infty} \sup | f_m'(x) - f_n'(x) | = 0.$$

Now, by Theorem 4.1.2 there exist continuous functions f and g such that

$$\lim_{n \to \infty} \sup | f_n(x) - f(x) | = 0$$

and

$$\lim_{n \to \infty} \sup | f_n'(x) - g(x) | = 0.$$

We have only to prove that $f'(x_0) = g(x_0)$ for $a \le x_0 \le b$. Now,

$$\begin{aligned}
f(x) - f(x_0) &= [f(x) - f_n(x)] + [f_n(x) - f_n(x_0)] \\
&\quad + [f_n(x_0) - f(x_0)] \\
&= [f(x) - f_n(x)] + f_n'(\xi)(x - x_0) \\
&\quad + [f_n(x_0) - f(x_0)]
\end{aligned}$$

(by Theorem 5.1.3), where ξ is between x and x_0. Taking the limit as $n \to \infty$,

$$f(x) - f(x_0) = g(\xi)(x - x_0) \qquad \text{or} \qquad \frac{f(x) - f(x_0)}{x - x_0} = g(\xi),$$

and

$$\lim_{x \to x_0} \frac{f(x) - f(x_0)}{x - x_0} = g(x_0). \quad \blacksquare$$

It should be emphasized that $C^{(1)}[a, b]$ is complete *under the norm of that space.* Thus, in order to conclude that the limit of a sequence of continuously differentiable functions has a continuous derivative, it must be determined not only that the sequence of functions converges uniformly, but also that the sequence of derivatives converges uniformly.

PROBLEMS

1. Let $f_n(x) = (\sin nx)/n$. Show $f_n \to 0$ uniformly, but that $\{f_n'\}$ converges only for integral multiples of 2π.

2. Let

$$f_n(x) = \begin{cases} x^{(1+1/n)} \sin \dfrac{1}{x} & \text{if } 0 < x \le \dfrac{4}{\pi}, \\[3mm] 0 & \text{if } x = 0, \end{cases}$$

and let $\lim_{n \to \infty} f_n(x) = f(x)$.

(a) Find $f(x)$.
(b) Find $f_n'(x)$, $x > 0$, and $f_n'(0+)$.
(c) Is f the limit of f_n in the sup norm?
(d) Show $f'(0+)$ does not exist.

In Problems 3–8,

(a) Find $f(x) = \lim_{n \to \infty} f_n(x)$.
(b) Determine whether $f_n \to f$ in the norm of $C^{(1)}[0, 1]$.
(c) Is $f \in C^{(1)}[0, 1]$?

(See the Appendix for graphs.)

3. $f_n(x) = -\dfrac{(e^{-nx} - n)}{n}$.

4. $f_n(x) = xe^{-nx}$.

5. $f_n(x) = \dfrac{1}{n} e^{-nx}$.

6. $f_n(x)$
$$= n\left[x(1 - x)^n + \dfrac{1}{n+1}\right].$$

7. $f_n(x) = x(e^{-nx} + x)$.

8. $f_n(x) = nxe^{-nx} + \dfrac{e^{x-1}}{n}$.

5.3 partial differentiation

If $u = f(x, y)$, it may be useful to consider the change in u caused by changing only x and y independently. Then u may have partial derivatives with respect to x or y defined by the following limits if they exist [The domain of definition must be some set in E^2 containing the rectangle from (x_0, y_0) to $(x_0 + h, y_0 + k)$.]

$$f_x(x_0, y_0) = \lim_{h \to 0} \frac{f(x_0 + h, y_0) - f(x_0, y_0)}{h},$$

$$f_y(x_0, y_0) = \lim_{k \to 0} \frac{f(x_0, y_0 + k) - f(x_0, y_0)}{k}.$$

Other symbols for the function f_x are $\partial f/\partial x$ and f_1.

The total change in u in terms of changes in x and y and their partial derivatives is basic in proving many formulas involving partial derivatives.

5.3.1 *THEOREM* (Fundamental increment formula) Let $u = f(x, y)$ have continuous partial derivatives at (x_0, y_0) and in some region that contains $\{(x, y) \mid a < x_0 - h < x_0 + h < b, c < y_0 - k < y_0 + k < d\}$. Let increments in x, y, and u be denoted by Δx, Δy, and Δu, respectively, and let $|\Delta x| < h$, $|\Delta y| < k$. Then,

$$\Delta u = \frac{\partial u}{\partial x} \Delta x + \frac{\partial u}{\partial y} \Delta y + \epsilon_1 \Delta x + \epsilon_2 \Delta y,$$

where $\epsilon_1, \epsilon_2 \to 0$ as $\Delta x, \Delta y \to 0$.

Proof By definition,

$$\Delta u = f(x_0 + \Delta x, y_0 + \Delta y) - f(x_0, y_0)$$

$$= f(x_0 + \Delta x, y_0 + \Delta y) - f(x_0, y_0 + \Delta y)$$

$$+ f(x_0, y_0 + \Delta y) - f(x_0, y_0).$$

Now, apply the mean value theorem for derivatives to both differences above, so

$$\Delta u = f_x(x_0 + \theta_1 \Delta x, y_0 + \Delta y)\ \Delta x + f_y(x_0, y_0 + \theta_2 \Delta y)\ \Delta y,$$

where $0 < \theta_1, \theta_2 < 1$. Now, f_x and f_y are continuous, so

$$f_x(x_0 + \theta_1 \Delta x, y_0 + \Delta y) = f_x(x_0, y_0) + \epsilon_1,$$

where $\epsilon_1 \to 0$ as $\Delta x, \Delta y \to 0$
and

$$f_y(x_0, y_0 + \theta_2 \Delta y) = f_y(x_0, y_0) + \epsilon_2,$$

where $\epsilon_2 \to 0$ as $\Delta x, \Delta y \to 0.$

Substituting in the formula for Δu given above and dropping subscripts to obtain a general formula,

$$\Delta u = f_x(x, y)\ \Delta x + f_y(x, y)\ \Delta y + \epsilon_1 \Delta x + \epsilon_2 \Delta y. \quad \blacksquare$$

This fundamental increment formula may be used to obtain any number of formulas under the general heading of "chain rules." For example, if $u = f(x, y)$, $x = x(s, t)$, $y = y(s, t)$, s and t are independent variables, x and y are intermediate variables, and u is the dependent variable, then

$$\frac{\partial u}{\partial s} = \frac{\partial u}{\partial x}\frac{\partial x}{\partial s} + \frac{\partial u}{\partial y}\frac{\partial y}{\partial s},$$

and a similar formula for $\partial u/\partial t$ holds.

Since f_x and f_y are functions of x and y, we may have higher-order derivatives, with obvious definitions, written

$$f_{xy}(x, y) = \frac{\partial^2 f}{\partial y\,\partial x}, \qquad f_{xx}(x, y) = \frac{\partial^2 f}{\partial x^2},$$

and so on.

It should be recalled that the fundamental increment formula is also the motivation for defining the total differential of $u = f(x, y)$ as

$$du = \frac{\partial f}{\partial x}\,dx + \frac{\partial f}{\partial y}\,dy.$$

PROBLEMS

1. Let $u = f(x, y)$, $x = x(s, t)$, $y = y(s, t)$, and let the partial derivatives of u with respect to x and y, of x with respect to s and t, and of y with respect to s and t all exist and be continuous. Prove

$$\frac{\partial u}{\partial s} = \frac{\partial u}{\partial x}\frac{\partial x}{\partial s} + \frac{\partial u}{\partial y}\frac{\partial y}{\partial s}.$$

2. Let $u = F(x, y) = 0$, and let $F_x(x, y)$ and $F_y(x, y)$ exist with $F_y(x, y) \neq 0$. Prove

$$\frac{dy}{dx} = -\frac{F_x(x, y)}{F_y(x, y)}.$$

3. Let $u = f(x, y), f_x, f_y, f_{xy}, f_{yx}$ all be continuous. Prove $f_{xy} = f_{yx}$.

    [*Hint*: let $\Delta = f(x_0 + \Delta x, y_0 + \Delta y) - f(x_0 + \Delta x, y_0) - f(x_0, y_0 + \Delta y) + f(x_0, y_0)$; let $\varphi(x, y) = f(x + \Delta x, y) - f(x, y)$ and $\psi(x, y) = f(x, y + \Delta y) - f(x, y)$; then

    $\Delta = \varphi(x_0, y_0 + \Delta y) - \varphi(x_0, y_0) = \psi(x_0 + \Delta x, y_0) - \psi(x_0, y_0)$.

    Apply the mean value theorem twice to both forms of Δ.]

4. If $u = f(x, y)$, $x = r \cos\theta$, $y = r \sin\theta$, and if the first and second partial derivatives of u with respect to x and y all exist and are continuous, show that

$$\frac{\partial^2 f}{\partial\theta^2} = r^2 \sin^2\theta\,\frac{\partial^2 f}{\partial x^2} + r^2 \cos^2\theta\,\frac{\partial^2 f}{\partial y^2} - 2r^2 \sin\theta\cos\theta\,\frac{\partial^2 f}{\partial y\,\partial x}$$

$$-r\cos\theta\,\frac{\partial f}{\partial x} - r\sin\theta\,\frac{\partial f}{\partial y}.$$

    (*Hint*: what is $\partial(\partial f/\partial x)/\partial\theta$? Note that $\partial f/\partial x$ is a function of the two intermediate variables, x and y, and θ is an independent variable.)

5. Find $\partial^2 f/\partial r^2$ in Problem 4.

6.   Let $u = f(x, y, z)$, where second-order partials exist and are continuous. Show that the Laplacian

$$\nabla^2 u = \frac{\partial^2 f}{\partial x^2} + \frac{\partial^2 f}{\partial y^2} + \frac{\partial^2 f}{\partial z^2}$$

becomes

$$\frac{\partial^2 f}{\partial r^2} + \frac{1}{r^2}\frac{\partial^2 f}{\partial \theta^2} + \frac{1}{r}\frac{\partial f}{\partial r} + \frac{\partial^2 f}{\partial z^2}$$

in cylindrical coordinates.

7.   A function $f(x, y)$ is homogeneous of degree n if

$$f(tx, ty) = t^n f(x, y), \qquad t > 0.$$

Euler's formula states that if $f(x, y)$ is homogeneous of degree n and has continuous partial derivatives, then

$$x\frac{\partial f}{\partial x} + y\frac{\partial f}{\partial y} = nf.$$

Verify Euler's formula for

(a) $x^3 - 3x^2y$,

(b) $\dfrac{xy}{x^2 + y^2}$,

(c) $\sqrt{x^2 + xy + y^2}$.

8.   Prove Euler's formula as follows: let $g(t) = f(u, v)$, where $u = tx$ and $v = ty$. Find $g'(t)$ by the chain rule and then let $t = 1$.

9.   A function $u(x, y)$ that satisfies Laplace's equation

$$\frac{\partial^2 u}{\partial x^2} + \frac{\partial^2 u}{\partial y^2} = 0$$

is said to be *harmonic*. Show that the following are harmonic:

(a) $u = \ln(x^2 + y^2)$,
(b) $u = e^x \sin y$,

(c) $u = \tan^{-1}\dfrac{y}{x}$.

10. Let $u = f(x, y)$ and $v = g(x, y)$ have continuous first and second partial derivatives. They are said to satisfy the Cauchy–Riemann conditions if

$$\frac{\partial u}{\partial x} = \frac{\partial v}{\partial y} \quad \text{and} \quad \frac{\partial u}{\partial y} = -\frac{\partial v}{\partial x}.$$

Show that if u and v satisy the Cauchy–Riemann conditions, they are harmonic.

11. Let

$$f(x, y) = \frac{xy}{x^2 + y^2}, \qquad (x, y) \neq (0, 0), \qquad f(0, 0) = 0.$$

(a) Show that f is not continuous at $(0, 0)$ in the sense of Section 4.4.
(b) Find $f_x(x, y)$, $(x, y) \neq (0, 0)$.
(c) Find

$$\lim_{y \to 0}(\lim_{x \to 0} f_x(x, y)) \quad \text{and} \quad \lim_{x \to 0}(\lim_{y \to 0} f_x(x, y)).$$

(d) Find $f_x(0, 0)$ from the definition.

12. Let

$$f(x, y) = (x^2 + y^2) \tan^{-1}\frac{y}{x}, \qquad (x, y) \neq (0, 0), \qquad f(0, 0) = 0.$$

(a) Does $\lim f(x, y)$ exist as $(x, y) \to (0, 0)$, in the sense of Section 4.4?
(b) Find $f_{xy}(x, y)$ and $f_{yx}(x, y)$, $(x, y) \neq (0, 0)$.
(c) In the definition of

$$f_x(x_0, y_0) = \lim_{h \to 0} \frac{f(x_0 + h, y_0) - f(x_0, y_0)}{h},$$

replace $f(x_0, y_0)$ by $\lim_{x \to x_0} f(x, y_0)$. With this definition of $f_x(x_0, y_0)$ and similar definitions for other partial derivatives, find $f_{xy}(0, 0)$ and $f_{yx}(0, 0)$.

5.4 Taylor's formula—analytic functions

An early use of differential calculus was the expansion of functions in power series which in many ways resemble finite polynomials. In Chapter VII several different techniques will be investigated and the resulting series examined for convergence properties, but for the present there is one general method that, theoretically, gives a power series expansion for every function that has an infinite number of derivatives. Two problems arise, however. First, the calculation of higher-order derivatives may be unusually difficult; and, second, the question of whether or not the infinite series obtained actually converges *to the function* may not be easy to answer. Nevertheless, Taylor's formula (and the series to which it gives rise) is one of the most important formulas in calculus.

5.4.1 *THEOREM* (Taylor's formula) Let f have n derivatives in some interval containing a and b as interior points. Then

$$f(b) = f(a) + f'(a)(b - a) + \frac{f''(a)}{2!}(b - a)^2$$

$$+ \frac{f'''(a)}{3!}(b - a)^3 + \cdots + \frac{f^{(n-1)}(a)}{(n - 1)!}(b - a)^{n-1}$$

$$+ \frac{f^{(n)}(\xi)}{n!}(b - a)^n, \qquad a < \xi < b.$$

Proof Although several proofs of this formula are known, none is well motivated. The following is brief and may not be too obscure when the possibility of "telescoping" series has been experienced. In this case it occurs after differentiating term by term.

Let K be defined by

$$f(b) = f(a) + f'(a)(b - a) + \frac{f''(a)}{2!}(b - a)^2$$

(equation continues)

$$+ \cdots + \frac{f^{(n-1)}(a)}{(n-1)!} (b-a)^{n-1} + \frac{K(b-a)^n}{n!}.$$

Now let

$$g(x) = f(b) - \left[f(x) + f'(x)(b-x) + \frac{f''(x)}{2!} (b-x)^2 \right.$$

$$\left. + \cdots + \frac{f^{(n-1)}(x)}{(n-1)!} (b-x)^{n-1} + \frac{K(b-x)^n}{n!} \right].$$

Obviously, $g(b) = 0$, and from the definition of K, $g(a) = 0$.
Term by term differentiation gives

$$g'(x) = - \frac{(b-x)^{n-1}}{(n-1)!} f^{(n)}(x) + \frac{K(b-x)^{n-1}}{(n-1)!}$$

Since the hypotheses of Rolle's theorem are satisfied, there is
a value ξ between x and b such that $g'(\xi) = 0$. Thus,
$K = f^{(n)}(\xi)$. Now, replacing x by a yields the desired
formula. ∎

Although in the above proof we wanted to differentiate with
respect to the left-hand end point, applications of the formula are
usually stated with the right-hand end point as a variable.
So,

$$f(x) = f(a) + f'(a)(x-a) + \frac{f''(a)}{2!} (x-a)^2$$

$$+ \cdots + \frac{f^{(n-1)}(a)}{(n-1)!} (x-a)^{n-1} + R_n,$$

where

$$R_n = \frac{f^{(n)}(\xi)}{n!} (x-a)^n, \qquad \text{where} \quad a < \xi < x.$$

The form for R_n as given here is called Lagrange's form of the remainder. If we write the above series as $f(x) = S_n + R_n$, it is obvious
that $\lim_{n \to \infty} S_n = f(x)$ if and only if $\lim_{n \to \infty} R_n = 0$. It is tempting

simply to test the series for convergence by the ratio test or the root
test, and, if it is discovered that the series converges for a certain
set of values, to conclude that the series *converges to* $f(x)$ for these
values. The unfortunate situation may exist, however, that the series
converges but not to $f(x)$.

5.4.2   EXAMPLE Let

$$f(x) = \begin{cases} e^{-1/x^2} & \text{if } x \neq 0, \\ 0 & \text{if } x = 0. \end{cases}$$

A little calculation shows that all derivatives are zero when
$x = 0$. Thus, Taylor's series certainly converges, but not to
$f(x)$.

5.4.3   EXAMPLE Let $f(x) = \sin x$, $a = 0$. Then, $f(0) = 0$,
$f'(0) = 1, f''(0) = 0, f'''(0) = -1$, and so on. So,

$$f(x) = x - \frac{x^3}{3!} + \frac{x^5}{5!} - \cdots + (-1)^{n+1}\frac{x^{2n-1}}{(2n-1)!} + R_{2n+1}.$$

(Note that we have omitted the terms that are zero and thus
have renumbered the terms in the series; that is, n no longer
gives the nth term in the original series.) Now

$$R_{2n+1} = \frac{f^{(2n+1)}(\xi)x^{2n+1}}{(2n+1)!},$$

where $f^{(2n+1)}(\xi)$ is $\cos \xi$, $-\sin \xi$, $-\cos \xi$, or $+\sin \xi$, and ξ is
between 0 and x. In any case,

$$|R_{2n+1}| \leq \frac{|x^{2n+1}|}{(2n+1)!} \to 0 \qquad \text{as} \quad n \to \infty \quad \text{for all} \quad x.$$

Therefore the series converges to $\sin x$.

If, as in the above example, $a = 0$, Taylor's series is called *Maclaurin's
series*.

We have seen in Theorem 5.1.1 that a differentiable function is
continuous. The converse is not true, however, as the example
$f(x) = |x|$ at $x = 0$ shows. In fact, examples have been given of

functions that are continuous everywhere and differentiable nowhere. In a sense they have corners like the one in $y = |x|$ at too many points to have a derivative anywhere.

The function

$$f(x) = \begin{cases} x \sin \dfrac{1}{x} & \text{if } x \neq 0, \\ \\ 0 & \text{if } x = 0, \end{cases}$$

is another example of a function that is continuous at zero but not differentiable there; while

$$f(x) = \begin{cases} x^2 \sin \dfrac{1}{x} & \text{if } x \neq 0, \\ \\ 0 & \text{if } x = 0. \end{cases}$$

has a derivative at zero but the derivative is not continuous there. In fact, by changing the power of x, the function

$$f(x) = \begin{cases} x^n \sin \dfrac{1}{x} & \text{if } x \neq 0, \\ \\ 0 & \text{if } x = 0, \end{cases}$$

can be made to have any number of derivatives desired with the last one either continuous or not at $x = 0$.

Finally, Example 5.4.2 shows a function that has an infinite number of continuous derivatives, but Taylor's series does not converge to the function.

In a sense the best behaved function is one that has an infinite number of derivatives in a region and for which Taylor's series converges to the function itself there. Such a function is said to be *analytic* in this region.

The student may wish to compare the situation described here, with varying stages of "good behavior," with the situation of a complex-valued function of a complex variable. In the latter case the existence of one derivative implies the existence of all of them and of the convergence of Taylor's series to the function.

PROBLEMS

1. Show that $f(x) = e^x$ is analytic for all x by expanding it in Maclaurin's series and showing $|R_n| \to 0$ for all x.

2. Expand $f(x) = \cos x$ in powers of x and show the remainder approaches zero as $n \to \infty$ for all x. If only three terms are used to approximate $\cos 0.1$, what is an upper bound for the error?

3. Show $f(x) = \sinh x$ is analytic for all x.

4. Find the derivative of

$$f(x) = \begin{cases} x^2 \sin \dfrac{1}{x} & \text{if } x \neq 0, \\[2mm] 0 & \text{if } x = 0. \end{cases}$$

(*Note*: take the cases of $x = 0$ and $x \neq 0$ separately.)

5. Find examples of a function with the following at $x = 0$ (Prove your assertions for the functions.):

(a) $f'(0)$ exists, $f'(x)$ is continuous at 0, but $f''(0)$ does not exist.

(b) $f''(0)$ exists, but $f''(x)$ is not continuous at $x = 0$.

6. Verify that all derivatives are zero at $x = 0$ in Example 5.4.2.

7. With the notation of Theorem 5.4.1, let

$$\Phi(x) = g(x) + \frac{1}{n!} K(b - x)^n.$$

Then,

$$\Phi(b) = 0, \qquad \Phi(a) = \frac{K}{n!} (b - a)^n,$$

and

$$\Phi(b) - \Phi(a) = \Phi'(\xi)(b - a),$$

by the mean value theorem. Use this to show that

$$R_n = \frac{(x - a)(x - \xi)^{n-1}}{(n - 1)!} f^{(n)}(\xi), \qquad \text{where} \quad a < \xi < x.$$

This is called *Cauchy's form of the remainder*. Note that in general ξ is different in the two different forms of the remainder.

8. Expand $\ln(1 + x)$ in Maclaurin's series. Show that Lagrange's form of the remainder approaches zero as $n \to \infty$ if $-\frac{1}{2} < x \le 1$ and that Cauchy's form of the remainder approaches zero as $n \to \infty$ if $-1 < x < 1$.

9. Expand x^4 in terms of $(x - 1)$.

10. Expand $\sin x$ in terms of $(x - \frac{1}{3}\pi)$. Find $\sin 61°$ correct to three decimal places.

VI

RIEMANN
INTEGRABLE
FUNCTIONS

6.1 the Riemann integral

The definition of the Riemann integral has a parallel that, with exactly the same notation, defines a certain area. Thus in our development the integral will exist if and only if the area does, and one may be used to find the other. Considered as an area this development is intuitive and suggests many properties of the integral even though the latter is essentially an analytical concept.

Let $f(x) \geq 0$ and bounded for $a \leq x \leq b$, and form a *partition P* of $[a, b]$: $a = x_0 < x_1 < x_2 < \cdots < x_n = b$. Let $\Delta x_i = x_i - x_{i-1}$ and $\Delta = \sup_i(x_i - x_{i-1})$. Then Δ is called the *norm* of the partition P (not to be confused with $\| \cdot \|$). Let $M_i = \sup f(x)$, $m_i = \inf f(x)$ for $x_{i-1} < x < x_i$, $i = 1, 2, \ldots, n$. Finally, let

$$\underline{s_n} = \sum_{i=1}^{n} m_i \, \Delta x_i, \qquad \overline{S_n} = \sum_{i=1}^{n} M_i \, \Delta x_i.$$

Then, if $\lim_{\Delta \to 0} \underline{s_n} = \lim_{\Delta \to 0} \overline{S_n}$, f is Riemann integrable on $[a, b]$. The common limit is called the *integral* and is denoted by

$$\int_a^b f(x) \ dx.$$

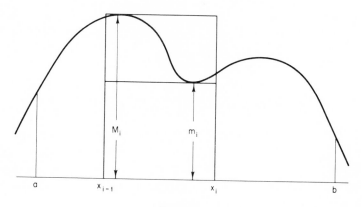

Figure 7

6.1.1   *REMARKS*

(a) Note that if $\int_a^b f(x) \ dx$ exists, M_i and m_i could be replaced by any value $f(\xi_i)$ such that $x_{i-1} \le \xi_i \le x_i$, for then $m_i \le f(\xi_i) \le M_i$. This gives

$$\int_a^b f(x) \ dx = \lim_{\Delta \to 0} \sum_{i=1}^n f(\xi_i) \ \Delta x_i.$$

(b) The restriction that $f(x) \ge 0$ is easily removed by letting

$$f^+(x) = \begin{cases} f(x) & \text{for all } x \text{ such that } f(x) \ge 0, \\ 0 & \text{elsewhere,} \end{cases}$$

$$f^-(x) = \begin{cases} -f(x) & \text{for all } x \text{ where } f(x) < 0, \\ 0 & \text{elsewhere,} \end{cases}$$

and noting that $f = f^+ - f^-$. Then define

$$\int_a^b f(x) \, dx = \int_a^b f^+(x) \, dx - \int_a^b f^-(x) \, dx$$

if and only if both integrals on the right-hand side exist. Since $|f(x)| = f^+(x) + f^-(x)$, a function is integrable only if its absolute value is integrable. This would not be the case if we simply allowed m_i and M_i to be negative in the definition of the integral. Henceforth theorems and definitions concerning integration will be understood to hold first for nonnegative functions and then be extended by the above convention to functions with both positive and negative values.

(c) As a consequence of the above,

$$\left| \int_a^b f(x) \, dx \right| \leq \int_a^b |f(x)| \, dx.$$

(d) We shall use the following conventions (which are equivalent to allowing $x_{i-1} \geq x_i$ in the definition).

$$\int_b^a f(x) \, dx = - \int_a^b f(x) \, dx \qquad \text{and} \qquad \int_a^a f(x) \, dx = 0.$$

(e) It should be emphasized that f is bounded on $[a, b]$ and that both a and b are finite. These restrictions will be relaxed in the discussion of improper integrals, and at that time it is important to note that the extension to functions with negative values as discussed in (b) is done *after* the domain and range are allowed to become infinite.

(f) It should be noted that $\overline{S_n}$ and $\underline{s_n}$ depend on both the function and the partition as well as on n. It is therefore sometimes useful to adopt a notation such as $\overline{S_n}(f, P)$ and $\underline{s_n}(f, P)$. Then, since $\overline{S_n}(f, P)$ and $\underline{s_n}(f, P)$ are monotone, $\lim_{\Delta \to 0} \overline{S_n}(f, P)$ and $\lim_{\Delta \to 0} \underline{s_n}(f, P)$ always exist (finite) if f is bounded. They are sometimes written

$$\overline{\int_a^b} f(x) \, dx \qquad \text{and} \qquad \underline{\int_a^b} f(x) \, dx,$$

respectively, and are called the *upper* and *lower Riemann integrals* of f on $[a, b]$.

One question that arises immediately is: What functions are integrable? A partial answer is given in the following theorem.

6.1.2   THEOREM   If f is continuous on $[a, b]$, then f is Riemann integrable on $[a, b]$.

   Proof   We shall show that $\lim_{\Delta \to 0}(\overline{S_n} - \underline{s_n}) = 0$. Fix $\epsilon > 0$. Then, $\overline{S_n} - \underline{s_n} = \sum_{i=1}^{n} (M_i - m_i)\,\Delta x_i$. Since f is continuous on $[a, b]$ it is uniformly continuous on $[a, b]$. Therefore, there is a $\delta > 0$ such that if $|\Delta x_i| < \delta$,

$$M_i - m_i < \frac{\epsilon}{b - a}.$$

Now let $\Delta < \delta$. Then,

$$\overline{S_n} - \underline{s_n} < \frac{\epsilon}{b - a} \sum_{i=1}^{n} \Delta x_i = \epsilon. \quad \blacksquare$$

Some familiar results from elementary calculus are stated here as theorems, and the proofs are left as problems. In many proofs involving more than one integral, it may happen that different partitions of the interval are used in defining m_i, M_i, $\underline{s_n}$, and $\overline{S_n}$ for the different integrals. In this case a simple device is to include the points of all partitions in a new "finer" partition applicable to all the integrals involved. The following lemma is then useful.

6.1.3   LEMMA   Let $a = x_0 < x_1 < x_2 < \cdots < x_{n-1} < x_n = b$ and $a = x_0' < x_1' < x_2' < \cdots < x_{m-1}' < x_m' = b$, $m > n$, be two partitions of $[a, b]$ with every x_i the same as some x_j' (thus the second partition is finer than the first). Let m_i, M_i, $\underline{s_n}$, and $\overline{S_n}$ be the inf, sup, lower sum, and upper sum for the first partition, and m_j', M_j', $\underline{s_m'}$, and $\overline{S_m'}$ similar quantities for the second partition. Then

$$\underline{s_n} \leq \underline{s_m'} \leq \overline{S_m'} \leq \overline{S_n}.$$

   Outline of Proof   Consider adding one point at a time to the first partition. $\quad \blacksquare$

6.1.4 *THEOREM* If f and g are integrable on $[a, b]$, and if $f(x) \leq g(x)$, then

$$\int_a^b f(x) \, dx \leq \int_a^b g(x) \, dx.$$

6.1.5 *THEOREM* If any two of the following integrals exist, the third integral exists, and

$$\int_a^b f(x) \, dx = \int_a^c f(x) \, dx + \int_c^b f(x) \, dx.$$

6.1.6 *THEOREM* If f and g are integrable on $[a, b]$, and α and β are constants, $\alpha f + \beta g$ is integrable on $[a, b]$, and

$$\int_a^b [\alpha f(x) + \beta g(x)] \, dx = \alpha \int_a^b f(x) \, dx + \beta \int_a^b g(x) \, dx.$$

In view of Theorem 6.1.6 the set of functions integrable on $[a, b]$ forms a linear space which we shall investigate in more detail shortly. For the present we can improve on Theorem 6.1.2 slightly by the following:

6.1.7 *THEOREM* If f is bounded on $[a, b]$ and continuous on $[a, b]$ except at a finite number of points, f is integrable on $[a, b]$.

Proof Let $|f(x)| < M$ on $[a, b]$ and suppose f is discontinuous only at c, between a and b. For $\epsilon > 0$, let δ be a positive number less than $\epsilon/4M$. Now, f is integrable on $[a, c - \delta]$ and $[c + \delta, b]$. Let $\underline{s}$ and $\bar{S}$ be the lower and upper sums for any partition of $[c - \delta, c + \delta]$. Then, no matter what the partition is, $\underline{s} \leq M2\delta < \epsilon/2$, and $\bar{S} \leq M2\delta \leq \epsilon/2$. Then,

$$|\bar{S} - \underline{s}| \leq \bar{S} + \underline{s} < \frac{\epsilon}{2} + \frac{\epsilon}{2} = \epsilon.$$

Thus f is integrable on $[c - \delta, c + \delta]$, and, by Theorem 6.1.5, integrable on $[a, b]$. Similarly, f may be discontinuous at a finite number of points, including a and b. ∎

PROBLEMS

1. Prove Lemma 6.1.3. Draw a figure to show what happens when one point is added to the partition P in the subinterval $[x_{i-1}, x_i]$.

2. Prove Theorem 6.1.4.

3. Prove Theorem 6.1.5. Note that this must hold for any order of the points a, b, and c.

4. Prove Theorem 6.1.6.

5. Calculate $\int_a^b x \, dx$ from the definition using both $\underline{s}_n$ and $\overline{S}_n$. (*Hint*: take equal subintervals; recall that $\sum_{i=1}^n i = n(n+1)/2$.)

6. Calculate $\int_a^b x^2 \, dx$ from the definition.

7. Determine whether the following integrals exist, according to criteria given in this section:

   (a) $\displaystyle\int_0^1 \sin \frac{1}{x} \, dx.$

   (b) $\displaystyle\int_0^1 f(x) \, dx,$ where $f(x) = \begin{cases} 1 & \text{if } x \text{ is irrational,} \\ 0 & \text{if } x \text{ is rational.} \end{cases}$

   (c) $\displaystyle\int_{-1}^{+1} \frac{e^x - 1}{x(x^2 - 4)} \, dx.$

   (d) $\displaystyle\int_{-1}^{+1} \frac{e^x - 1}{x(x^2 - 1)} \, dx.$

8. Find $\int_0^{20} [x] \, dx$, where $[x]$ is the greatest integer less than or equal to x.

9. Without using antiderivatives, find $\int_{-2}^3 |x| \, dx$.

10. Give an example of a function f such that

$$\int_0^1 |f(x)| \, dx \quad \text{exists,} \quad \text{but} \quad \int_0^1 f(x) \, dx \quad \text{does not.}$$

In Problems 11–13 divide the interval $[0, 1]$ into n equal parts by the partition $x_0 = 0, x_1 = 1/n, \ldots, x_n = n/n = 1$. (If the integrand does not exist for some value of x, define it so as to make it continuous.) Write out several terms, including the first and last, in the sums $\underline{s_n}$ and $\overline{S_n}$.

11. $\displaystyle\int_0^1 \frac{\sin x}{x+1}\,dx.$

12. $\displaystyle\int_0^1 e^{-1/x^2}\,dx.$

13. $\displaystyle\int_0^1 \frac{e^x - 1}{x}\,dx.$

In Problems 14–16, sums of the form $\underline{s_n}$ or $\overline{S_n}$ are given. Set up the corresponding integral with proper limits.

14. $\displaystyle\frac{1}{n}\left(\ln\frac{1}{n} + \ln\frac{2}{n} + \cdots + \ln\frac{n}{n}\right).$

15. $\displaystyle\frac{1}{n+1} + \frac{1}{n+2} + \frac{1}{n+3} + \cdots + \frac{1}{n+n}.$

16. $\displaystyle\frac{n}{1+n^2} + \frac{n}{4+n^2} + \cdots + \frac{n}{n^2+n^2}.$

6.2 antiderivatives—differentiation of the integral

Corresponding to the mean value theorem for derivatives there is a similar mean value theorem for integrals. It is the same sort of theorem in that it guarantees the existence of a point rather than describing a method of finding some point.

6.2.1   THEOREM   Let f be continuous on $[a, b]$. Then there is a value ξ between a and b such that

$$\int_a^b f(x)\,dx = f(\xi)\,(b - a).$$

Proof Since f is continuous, $\int_a^b f(x)\ dx$ exists, and, from the definition,

$$m(b-a) \le \int_a^b f(x)\ dx \le M(b-a),$$

where $m = \inf_{a \le x \le b} f(x)$, $M = \sup_{a \le x \le b} f(x)$. Then

$$m \le \frac{\displaystyle\int_a^b f(x)\ dx}{b-a} \le M.$$

By Theorem 4.2.4, there is a value ξ between a and b such that

$$f(\xi) = \frac{\displaystyle\int_a^b f(x)\ dx}{b-a}$$

or

$$\int_a^b f(x)\ dx = (b-a)f(\xi). \quad \blacksquare$$

6.2.2 *THEOREM* Let f be integrable on $[a,\ b]$ and continuous in some subinterval including x_0 between a and b. Let $\phi(x) = \int_a^x f(t)\ dt$. Then ϕ is differentiable at x_0 and $\phi'(x_0) = f(x_0)$.

Proof Let h be such that $x_0 + h$ is in the subinterval where f is continuous. Then

$$\phi(x_0 + h) - \phi(x_0) = \int_a^{x_0+h} f(x)\ dx - \int_a^{x_0} f(x)\ dx$$

$$= \int_{x_0}^{x_0+h} f(x)\ dx = f(\xi) \cdot h,$$

where ξ is between x_0 and $x_0 + h$, by Theorem 6.2.1. Then, since f is continuous,

$$\lim_{h \to 0} \frac{\phi(x_0 + h) - \phi(x_0)}{h} = f(x_0). \quad \blacksquare$$

6.2.3 **THEOREM** Let f be continuous on $[a, b]$, and let F be such that $F'(x) = f(x)$ for $a \le x \le b$. Then

$$\int_a^b f(x)\ dx = F(b) - F(a).$$

Proof Let $\phi(x) = \int_a^x f(t)\ dt$, so by Theorem 6.2.2, $\phi'(x) = f(x)$. Therefore, $F'(x) = \phi'(x)$, so $\phi(x) - F(x) = C$ (constant). Since $\phi(a) = 0$, $F(a) = -C$. Then, $\int_a^b f(t)\ dt = F(b) - F(a)$, or, since t is a dummy variable, $\int_a^b f(x)\ dx = F(b) - F(a)$. $\blacksquare$

The functions F and ϕ used here are called *antiderivatives* or *primitives*.

Theorems 6.2.2 and 6.2.3 together comprise what is generally called the "fundamental theorem of integral calculus." For continuous, bounded functions on a finite, closed interval, little more need be said—we know the integral exists, and, if the primitive can be found in a satisfactory form, we can find the value of the integral. As we shall see, however, many important problems in the study of integration require extending the domain, the range, and the rule that constitute the function to be integrated.

First, however, consider an extension of Theorem 6.2.2 so that the upper limit of integration is a differentiable function of x.

6.2.4 **THEOREM** If u is a differentiable function of x and f is continuous on $[a, b]$ with $a \le u(x) \le b$, then $F(x) = \int_a^{u(x)} f(t)\ dt$ is differentiable with respect to x and

$$F'(x) = f\big(u(x)\big)u'(x).$$

Proof This is an immediate consequence Theorem 6.2.2 and the chain rule for differentiating a function of a function. $\blacksquare$

6.2.5 **COROLLARY** With the same hypotheses as above but with $F(x) = \int_{u(x)}^b f(t)\ dt$,

$$F'(x) = -f\big(u(x)\big)u'(x).$$

6.2.6 THEOREM Let $f(t, x)$ and $\partial f/\partial x$ be continuous for $a \le t \le b$ and $c \le x \le d$. Then, if $F(x) = \int_a^b f(t, x)\, dt$,

$$F'(x) = \int_a^b \frac{\partial}{\partial x} f(t, x)\, dt.$$

Proof Let h be such that $c \le x + h \le d$. Then

$$F(x + h) - F(x) = \int_a^b f(t, x + h)\, dt - \int_a^b f(t, x)\, dt$$

$$= \int_a^b \big[f(t, x + h) - f(t, x)\big]\, dt$$

$$= \int_a^b \frac{\partial}{\partial x} f(t, \xi) \cdot h\, dt,$$

where ξ is between x and $x + h$, by the mean value theorem for derivatives. Then

$$\frac{F(x + h) - F(x)}{h} = \int_a^b \frac{\partial}{\partial x} f(t, \xi)\, dt.$$

We claim that the limit of this as $h \to 0$ is

$$\int_a^b \frac{\partial}{\partial x} f(t, x)\, dt.$$

First note that an equivalent statement is

$$\lim_{h \to 0} \int_a^b \left| \frac{\partial}{\partial x} f(t, \xi) - \frac{\partial}{\partial x} f(t, x) \right| dt = 0.$$

To show the latter, fix $\epsilon > 0$ and choose a uniform δ, independent of t and x,[†] such that

---

[†] The δ chosen must be independent of t and x. That this is possible depends on *uniform* continuity of $\partial f/\partial x$ for $a \le t \le b$, $c \le x \le d$. The development of uniform continuity of a function of two variables in a closed rectangle proceeds the same as that for a function of one variable in a closed interval.

$$\left| \frac{\partial}{\partial x} f(t, \xi) - \frac{\partial}{\partial x} f(t, x) \right| < \frac{\epsilon}{b - a} \qquad \text{when} \quad 0 < |h| < \delta.$$

Then

$$\int_a^b \left| \frac{\partial}{\partial x} f(t, \xi) - \frac{\partial}{\partial x} f(t, x) \right| dt < \epsilon. \quad \blacksquare$$

6.2.7 *THEOREM* Under the assumptions in the above three theorems, if $F(x) = \int_{v(x)}^{u(x)} f(t, x)\, dt$,

$$F'(x) = \int_{v(x)}^{u(x)} \frac{\partial}{\partial x} f(t, x)\, dt$$

$$+ f\big(u(x), x\big) \cdot u'(x) - f\big(v(x), x\big) \cdot v'(x).$$

Proof The above proofs may be combined, or different variables may be introduced so that the combined result follows from an extended version of the chain rule. $\quad \blacksquare$

PROBLEMS

In Problems 1–4, find $F'(x)$. If the integrand is not defined for certain values of t in the interval of integration, let it be defined so as to make it continuous.

1. $F(x) = \displaystyle\int_x^{x^2} e^{-t^2}\, dt.$ **3.** $F(x) = \displaystyle\int_{x^2}^{x^4} \sin \sqrt{t}\, dt.$

2. $F(x) = \displaystyle\int_0^x \frac{\sin tx}{t}\, dt.$ **4.** $F(x) = \displaystyle\int_0^x \frac{e^{xt} - e^{-xt}}{t}\, dt.$

5. Let $F(x) = \displaystyle\int_x^{x^2} \frac{(xt + 1)^2}{t}\, dt.$

(a) Find $F'(x)$ by Theorem 6.2.7.

(b) Find $F'(x)$ by first using Theorem 6.2.3 and then differentiating.

6. If $y(x) = 4 \int_0^x (t - x)y(t)\, dt - \int_0^x (t - x)f(t)\, dt$, show that

$$\frac{d^2y}{dx^2} + 4y = f(x).$$

7. Let

$$s(n) = \frac{1}{n + 1} + \frac{1}{n + 2} + \cdots + \frac{1}{n + n}.$$

Find $\lim_{n \to \infty} s(n)$ by evaluating the definite integral called for in Problem 15, Section 6.1.

8. Let $s(n)$ be the series in Problem 16, Section 6.1. Find $\lim_{n \to \infty} s(n)$.

9. Let

$$s(n) = \frac{1}{n^2} \sum_{k=1}^{n} \sqrt{n^2 - k^2}.$$

Show that $\lim_{n \to \infty} s(n) = \pi/4$.

10. Evaluate $\int_1^x t^n\, dt$, $n \neq -1$, by Theorem 6.2.3, and then find the limit as $n \to -1$ by l'Hospital's rule.

11. State and prove the Heine–Borel theorem for E^2.

12. Using Problem 11, justify the footnote to Theorem 6.2.6 by proving a theorem analogous to Theorem 4.1.4.

6.3 improper integrals

If the restrictions that the domain and the range of a function be finite are relaxed, the integral must be defined by a limiting process. The main question that concerns us here is to what extent the "infinite parts" that occur in defining such an integral may be allowed to cancel each other. Thus

(a) Do we want $\int_{-\infty}^{+\infty} x\, dx$ to exist and equal 0 because $\lim_{R \to \infty} \int_{-R}^{R} x\, dx = 0$?

(b) Do we want $\int_{-1}^{+1} dx/x^2$ to exist and equal -2 even though $\int_0^1 dx/x^2$ is not finite by any reasonable definition?

(c) Do we want $\int_0^\infty (1/x) \sin x \, dx$ to exist and equal $\pi/2$ even though $\int_0^\infty (1/x) \,|\sin x|\, dx = \infty$?

The geometric interpretations of these questions (see Problem 1) give an indication of what "infinite parts" may conceivably be allowed to cancel each other.

The answers to these questions depend on what conventions are adopted regarding an infinite domain and an infinite range. Briefly, no reasonable convention for infinite integrals allows (b) to converge; the Cauchy principal value of an integral assigns finite values to (a) and (c); the improper integral studied in elementary calculus gives a finite value to (c) but not to (a); according to most modern theories of integration all three integrals diverge. Following the conventions adopted in Remark 6.1.1 (b) we shall be led to the latter situation. [It should not be inferred, however, that no integral with infinite domain or infinite range for the integrand can have a finite value; for example,

$$\int_0^1 \frac{dx}{\sqrt{x}} = 2 \quad\text{and}\quad \int_1^\infty \frac{dx}{x^2} = 1.$$

Also, other conventions have important uses. Thus, we shall say that the Cauchy principle value of the integral in (a) is zero and that the integral in (c) converges conditionally as an improper integral to $\pi/2$.]

The following definition is made with the intent of allowing *no* infinite parts to cancel each other.

6.3.1 *DEFINITION* Let $f(x) \geq 0$. (If $f(x) < 0$ for some values of x, follow Remark 6.1.1 *after* using the following four rules for f^+ and f^-.)

(a) If $\int_a^x f(t) \, dt$ exists for every $x > a$, then

$$\int_a^\infty f(t) \, dt = \lim_{x \to +\infty} \int_a^x f(t) \, dt$$

if this limit exists as a finite number.

(b) If $\int_x^b f(t)\,dt$ exists for every $x < b$, then

$$\int_{-\infty}^b f(t)\,dt = \lim_{x \to -\infty} \int_x^b f(t)\,dt$$

if this limit exists as a finite number.

(c) If $\int_a^x f(t)\,dt$ exists for $a < x < b$, and if $\lim_{t \to b-} f(t) = \infty$, then

$$\int_a^b f(t)\,dt = \lim_{x \to b-} \int_a^x f(t)\,dt,$$

if this limit exists as a finite number.

(d) If $\int_x^b f(t)\,dt$ exists for $a < x < b$, and if $\lim_{t \to a+} f(t) = \infty$, then

$$\int_a^b f(t)\,dt = \lim_{x \to a+} \int_x^b f(t)\,dt,$$

if this limit exists as a finite number.

If an integral falls into more than one of the above categories, each part must be considered separately. (For example, $\int_{-\infty}^{+\infty} dx/x^2$ falls into four categories.) If any of the above limits fails to exist as a finite number, the integral is said to *diverge*.

Since we must apply Remark 6.1.1 (b) in case $f(x) < 0$ for some values of x, $\int f(x)\,dx$ will exist only if $\int |f(x)|\,dx$ exists just as in the case of finite range and finite domain for the integrand.

An important method of evaluating integrals in the four categories above is found in the study of complex variables when contour integration is considered.[†] In this case, the Cauchy principle value is the result, and care must be taken to see whether the integral actually converges in the sense of this section.

6.3.2 EXAMPLE Let

$$f(x) = \frac{d}{dx}\left(x^2 \sin \frac{1}{x^2}\right)$$

† See Hille, E., "Analytic Function Theory", Vol. 1, p. 176. Ginn (Blaisdell), Boston, Massachusetts, 1959.

$$f(x) = 2x \sin \frac{1}{x^2} - \frac{2}{x} \cos \frac{1}{x^2}$$

and consider $\int_0^1 f(x) \, dx$. This will exist only if $\int_0^1 |f(x)| \, dx$ exists. But,

$$|f(x)| \geq \frac{2}{x} \left| \cos \frac{1}{x^2} \right| - 2x \geq \frac{1}{x} - 2x$$

for $[(2n + \tfrac{1}{3})\pi]^{-1/2} \leq x \leq [(2n - \tfrac{1}{3})\pi]^{-1/2}$ and it is easily seen that

$$\int_0^1 \left(\frac{1}{x} - 2x \right) dx$$

diverges.

If this seems to contradict the fundamental relationship between integrals and antiderivatives, note that $f(x)$ does not satisfy the hypotheses of Theorems 6.2.2 and 6.2.3. Also, it must be emphasized that all the theorems in Section 6.2 refer to bounded functions on finite intervals. When either of these conditions is changed, the problem may develop complications.

We shall not attempt here to extend the results of Section 6.2 to integrals of the types presented in this section. Individual problems can usually be considered on their own merits without an extensive theoretical background. Instead, we shall discuss two functions defied by integrals of this type.

6.3.3 *EXAMPLE* The gamma function is defined by

$$\Gamma(x) = \int_0^\infty t^{x-1} e^{-t} \, dt.$$

To find values of x for which this converges we must consider two cases in Definition 6.3.1, say, $\int_0^1$ and $\int_1^\infty$. Let $\delta > 0$. Then

$$\int_\delta^1 t^{x-1} e^{-t} \, dt < \int_\delta^1 t^{x-1} \, dt = \frac{1}{x} - \frac{\delta^x}{x}$$

which has a finite limit as $\delta \to 0$ if $x > 0$. Also, for a fixed

$x > 0$, there is an N such that

$$\int_N^b t^{x-1}e^{-t}\,dt < \int_N^b \frac{1}{t^2}\,dt = -\frac{1}{b} + \frac{1}{N}$$

which approaches $1/N$ as $b \to \infty$. Obviously the integral from 1 to N causes no trouble. Thus $\Gamma(x)$ is defined for $x > 0$. It is easy to see that if x is zero, the first integral diverges. For some properties of the gamma function, see the problems.

6.3.4   *EXAMPLE* Let f be continuous for $t \geq 0$ except at possibly a finite number of points. Then

$$\int_0^b e^{-st}f(t)\,dt$$

exists. Further, if $|\,e^{-s_0 t}f(t)\,| \leq M$ (constant) for $0 \leq t < \infty$, $F(s) = \int_0^\infty e^{-st}f(t)\,dt$ exists for $s > s_0$ and $F(s)$ is called the *Laplace transform of* $f(t)$. We write $\mathcal{L}(f) = F$.

PROBLEMS

1.   Draw the graphs of

(a) $y = x,\ -\infty < x < \infty$,

(b) $y = 1/x^2,\ -1 \leq x \leq +1$, and

(c) $y = (1/x)\sin x,\ 0 \leq x < \infty$,

and interpret the integrals described in this section as areas.

In Problems 2–4, determine whether the integral converges and if so, evaluate it.

2. $\displaystyle\int_{-\infty}^{+\infty} \frac{dx}{1 + x^2}.$

3. $\displaystyle\int_0^4 \frac{dx}{x^2 - 4x + 4}.$

4. $\displaystyle\int_0^1 \ln x \, dx.$

5.   Determine p so that $\int_0^1 dx/x^p$ converges.

6.   Determine p so that $\int_1^\infty dx/x^p$ converges.

7.   Derive Wallis' formulas

$$\int_0^1 \frac{x^n \, dx}{\sqrt{1 - x^2}} = \int_0^{\pi/2} \sin^n x \, dx = \int_0^{\pi/2} \cos^n x \, dx$$

$$= \begin{cases} \dfrac{2\cdot4\cdot6\cdots(n-1)}{3\cdot5\cdot7\cdots\ \ n} & \text{if } n \text{ is an odd} \\ & \text{integer greater than 1,} \\[2ex] \dfrac{1\cdot3\cdot5\cdots(n-1)\pi}{2\cdot4\cdot6\cdots\ \ n\ \ 2} & \text{if } n \text{ is an even} \\ & \text{integer greater than 1.} \end{cases}$$

(*Hint*: first use a substitution to show all three integrals are equal. Then use integration by parts and induction.)

In Problems 8–12 determine convergence or divergence of the integral according to Definition 6.3.1. Do not evaluate.

8. $\displaystyle\int_1^\infty \frac{1}{x^2} \sin x \, dx.$

11. $\displaystyle\int_0^\infty \frac{\sin x}{x^{3/2}} \, dx.$

9. $\displaystyle\int_0^1 \frac{\sin x}{\sqrt{x}} \, dx.$

12. $\displaystyle\int_0^\infty \frac{\cos x}{\sqrt{1 + x^3}} \, dx.$

10. $\displaystyle\int_{-\infty}^{+\infty} e^{-x^2} \, dx.$

13.   Let $s(n)$ be the series in Problem 14, Section 6.1. Evaluate an improper integral by methods of this section to find $\lim_{n \to \infty} s(n)$.

14.   If $x > 0$, use integration by parts to show that $\Gamma(x + 1) = x\Gamma(x)$.

15.   If $x > 0$, prove $\Gamma'(x) = \int_0^\infty t^{x-1}e^{-t} \ln t \, dt$ by the *method* (not the result, which does not apply in this case) of Theorem 6.2.6.

16.   Discuss the graph of $y = \Gamma(x)$, $x > 0$. Find $\lim_{x \to 0+} \Gamma(x)$ and $\lim_{x \to \infty} \Gamma(x)$.

17.   Show $\Gamma(x)$ is continuous. How can Problem 14 be used to define $\Gamma(x)$ for certain negative values?

18.   From the definition set up $\Gamma(\frac{1}{2})$ and reduce this to a familiar integral by the substitution $y = \sqrt{t}$.

19.   Obtain other forms of the gamma function by the substitutions (a) $t = y^2$, and (b) $t = ky$.

20.   Prove that the Laplace transform is linear.

21.   Show that $\mathcal{L}(f') = s\mathcal{L}(f) - f(0)$.

In Problems 22–26 find the Laplace transform and the values of s for which it is valid.

22. $\mathcal{L}(1)$.   24. $\mathcal{L}(\cos at)$.

23. $\mathcal{L}(\sin at)$.   25. $\mathcal{L}(e^{at})$.

26. $\mathcal{L}(t^n)$. For n a nonnegative integer, express the result in terms of $n!$ For other nonnegative n use the gamma function.

6.4 convergence problems

Since the integrable functions form a linear space, it is natural to ask what kind of norm may be imposed and what convergence properties hold. Before attempting an answer let us examine some of the difficulties involved by considering two examples:

6.4.1 EXAMPLE   Let

$$
f_n(x) = \begin{cases} n^2 x, & \text{if} \quad 0 \le x \le \dfrac{1}{n}, \\[2mm] -n^2 x + 2n, & \text{if} \quad \dfrac{1}{n} < x \le \dfrac{2}{n}, \\[2mm] 0, & \text{if} \quad \dfrac{2}{n} < x \le 1. \end{cases}
$$

Here, $\lim_{n \to \infty} f_n(x) = f(x) = 0$, but

$$\int_0^1 f_n(x) \, dx = 1,$$

which may be found from a figure or by integrating from 0 to $1/n$ and from $1/n$ to $2/n$.

6.4.2 *EXAMPLE* Let $\{r_n\}$ be the set of rational numbers in $[0, 1]$. (Since the rationals are countable, the subset in $[0, 1]$ is countable; thus the notation $r_1, r_2, \ldots, r_n, \ldots$ is justified.)
Let

$$f_n(x) = \begin{cases} 1 & \text{if } x = r_1, r_2, \ldots, r_n, \\ 0 & \text{elsewhere.} \end{cases}$$

Then, for each n, $\int_0^1 f_n(x) \, dx$ exists and equals zero. Also,

$$\lim_{n \to \infty} f_n(x) = f(x) = \begin{cases} 1 & \text{if } x \text{ is rational,} \\ 0 & \text{if } x \text{ is irrational.} \end{cases}$$

But, $f(x)$ is not Riemann integrable ($\bar{S}_n = 1$, $\underline{s}_n = 0$ for all partitions).

It might be suggested that some sort of uniform boundedness ($|f_n(x)| \leq M$ for all n) would take care of the problem in Example 6.4.1, but it would not help in Example 6.4.2. It turns out (see Theorem 6.4.4) that requiring uniform convergence does obviate the difficulties in both examples already given, but this is sometimes too strict a requirement as the following example shows:

6.4.3 *EXAMPLE* Let $f_n(x) = nx(1 - x)^n$, $0 \leq x \leq 1$. Again,

$$\lim_{n \to \infty} f_n(x) = f(x) = 0, \qquad 0 \leq x \leq 1$$

and the convergence is *not* uniform since

$$\lim_{n \to \infty} \sup_{0 \leq x \leq 1} |f_n(x) - f(x)| = \frac{1}{e}.$$

However, the sequence of integrals has all the convergence

properties we would like; that is, $\int_0^1 f(x)\ dx$ exists and equals $\lim_{n \to 0} \int_0^1 nx(1 - x)^n\ dx$. (Use integration by parts to evaluate $\int_0^1 nx(1 - x)^n\ dx$.)

Thus, it should be hoped that something less restrictive than uniform convergence could be imposed. Clearly, however, a norm defined in terms of the integral itself will not make the space of integrable functions complete, even if we require boundedness, in view of Example 6.4.2.

It is apparent that we are just picking at the problem instead of offering a comprehensive solution. For this, the study of Lebesgue integration is necessary where the space of integrable functions *is* complete under a norm defined in terms of the integral; and under reasonable conditions of boundedness, $\lim_{n \to \infty} \int f_n = \int \lim_{n \to \infty} f_n$.

We conclude with a theorem whose importance should not be minimized by the fact that its application is not so universal as we would like.

6.4.4   THEOREM   Let f_n be Riemann integrable on $[a, b]$, and let $\lim_{n \to \infty} f_n(x) = f(x)$ *uniformly* on $[a, b]$. Then f is integrable on $[a, b]$, and

$$\lim_{n \to \infty} \int_a^b f_n(x)\ dx = \int_a^b f(x)\ dx.$$

Proof   Fix $\epsilon > 0$. There exists an N such that if $k > N$,

$$|f_k(x) - f(x)| < \frac{\epsilon}{3(b - a)}. \qquad (*)$$

Fix that k. Now, pick a partition

$$a = x_0 < x_1 < x_2 < \cdots < x_n = b$$

of $[a, b]$ with $\Delta = \max \Delta x_i$ such that if $\Delta < \delta$ and if $\overline{S_n}(f_k, \Delta)$ and $\underline{s_n}(f_k, \Delta)$ are upper and lower sums for f_k,

$$\left(\overline{S_n}(f_k, \Delta) - \underline{s_n}(f_k, \Delta)\right) < \frac{\epsilon}{3}.$$

Let $\overline{S_n}(f, \Delta)$ and $\underline{s_n}(f, \Delta)$ be upper and lower sums for f with $\Delta < \delta$. (In view of Lemma 6.1.3 we may assume without

loss of generality that it is the same partition as that for f_k.)
Then the inequality (*) holds for the maximum and minimum
values of f_k and f in each subinterval in the above partition,
so

$$| \overline{S}_n(f, \Delta) - \underline{s}_n(f, \Delta) | \leq | \overline{S}_n(f, \Delta) - \overline{S}_n(f_k, \Delta) |$$

$$+ | \overline{S}_n(f_k, \Delta) - \underline{s}_n(f_k, \Delta) |$$

$$+ | \underline{s}_n(f_k, \Delta) - \underline{s}_n(f, \Delta) |$$

$$\leq \frac{\epsilon}{3(b-a)} \sum_{i=1}^{n} \Delta x_i + \frac{\epsilon}{3}$$

$$+ \frac{\epsilon}{3(b-a)} \sum_{i=1}^{n} \Delta x_i = \epsilon.$$

Thus f is integrable and so are $f - f_n$ and $| f - f_n |$ for $n = 1, 2, \ldots$. For ϵ and k as above,

$$\left| \int_a^b f(x)\, dx - \int_a^b f_k(x)\, dx \right| \leq \int_a^b | f(x) - f_k(x) |\, dx \leq \frac{\epsilon}{3},$$

which completes the proof.   ∎

PROBLEMS

1. Let $f_n(x) = n^2 x (1 - x)^n, 0 \leq x \leq 1$.

   (a) Find $\int_0^1 f_n(x)\, dx$.
   (b) Find $f(x) = \lim_{n \to \infty} f_n(x)$.
   (c) Is the convergence in (b) uniform?
   (d) Does $\int_0^1 f(x)\, dx = \lim_{n \to \infty} \int_0^1 f_n(x)\, dx$?

2. Answer parts (a)–(d) of Problem 1 for

$$f_n(x) = n^3 x (1 - x)^n.$$

3. Answer parts (a)–(d) of Problem 1 for

$$f_n(x) = nx e^{-nx}.$$

4. Answer parts (a)–(d) of Problem 1 for

$$f_n(x) = x^n + x^2.$$

5. Answer parts (a)–(d) of Problem 1 for

$$f_n(x) = n^2 x^n (1 - x) + x.$$

6. Review the formulation of *uniform boundedness* in Problem 16, Section 3.2. In which of the Problems 1–5 is the sequence $\{f_n\}$ uniformly bounded?

7. For the space $C[a, b]$ let

$$\|f\|_I = \int_a^b |f(x)| \, dx.$$

(a) Is this a norm in the sense of Definition 1.2.1?
(b) Construct an example of a sequence of functions $\{f_n\}$ with the pointwise limit f such that

(i) $f_n \in C[a, b]$,
(ii) $\|f_n - f\|_I \to 0$,
(iii) $f \notin C[a, b]$.

8. The series $\sum_{n=1}^\infty f_n(x)$ is said to be *boundedly convergent* on $[a, b]$ if it converges for every $x \in [a, b]$ and if

$$s_n(x) = \sum_{k=1}^n f_k(x) \le M \qquad \text{for every } x \in [a, b].$$

Prove that if a series is uniformly convergent on $[a, b]$ except in a neighborhood of a finite number of points and also boundedly convergent on $[a, b]$, then it may be integrated term by term and the integral of the sum is the sum of the integrals. (*Hint:* consider

$$\left| \int_a^b [s(x) - s_n(x)] \, dx \right|$$

and break the integral at one point as in Theorem 6.1.7.)

VII

INFINITE
SERIES OF
FUNCTIONS

7.1 functions expressed as infinite series

We have seen that the convergence of an infinite series is determined by the convergence of the infinite sequence of partial sums. If

$$s_n(x) = \sum_{k=1}^{n} f_k(x),$$

then

$$f_1(x) = s_1(x) \quad \text{and} \quad f_n(x) = s_n(x) - s_{n-1}(x) \quad \text{for} \quad n = 2, 3, \ldots.$$

So, convergence properties for sequences yield similar properties for series, and vice versa. Since we have used sequences in the discussions of completeness properties of various spaces, many results are available to be applied to the study of infinite series. The following definition is suggested by Definition 3.2.4:

7.1.1 *DEFINITION* Let $\{f_k\}$ be a sequence of functions with respective domains D_k. Let $s_n(x) = \sum_{k=1}^{n} f_k(x)$, $D = \cap D_k$, and $s(x) = \lim_{n \to \infty} s_n(x)$ for $x \in D$. Then, $\sum_{k=1}^{\infty} f_k$ converges *uniformly* to s for $x \in D$ if the sequence $\{s_n\}$ converges uniformly to s for $x \in D$, that is, if given $\epsilon > 0$ there exists an N such that if $n > N$, $|s_n(x) - s(x)| < \epsilon$.

A simple and direct test for uniform convergence is the following Weierstrass M-test. It also illustrates the essential feature of uniform convergence, the fact that N is independent of x.

7.1.2 *THEOREM* Let f_k be defined for $x \in D_k$ and let $D = \cap D_k$. If $\sum_{k=1}^{\infty} M_k$ is a convergent series of nonnegative constants, and if for every $x \in D$, $|f_k(x)| \leq M_k$, $k = 1, 2, \ldots$, then $\sum_{k=1}^{\infty} f_k(x)$ converges uniformly on D.

Proof By the comparison test for convergence of series with positive terms, $\sum_{k=1}^{\infty} |f_k(x)|$ converges, and then so does $\sum_{k=1}^{\infty} f_k(x)$. Let $s_n(x) = \sum_{k=1}^{n} f_k(x)$, and $s(x) = \lim_{n \to \infty} s_n(x)$. This pointwise limit exists, and we have only to show uniformity. Let $M = \sum_{k=1}^{\infty} M_k$, and fix $\epsilon > 0$. Then there exists an N such that if $n > N$, then $|\sum_{k=1}^{n} M_k - M| < \epsilon$. Now, for $n > N$,

$$|s_n(x) - s(x)| = |f_{n+1}(x) + f_{n+2}(x) + \cdots|$$
$$\leq |f_{n+1}(x)| + |f_{n+2}(x)| + \cdots$$
$$\leq M_{n+1} + M_{n+2} + \cdots$$
$$= \left| \sum_{k=1}^{n} M_k - M \right| < \epsilon. \quad \blacksquare$$

7.1.3 *THEOREM* Let $\sum_{k=0}^{\infty} c_k x^k$ converge for $|x| \leq r$, or just for $|x| < r$. Then $\sum_{k=0}^{\infty} c_k x^k$ converges *uniformly* for $|x| \leq r_1 < r$.

Proof Pick t such that $r_1 < t < r$. Then, since $\sum_{k=0}^{\infty} c_k t^k$ converges, there is an A such that $|c_k t^k| < A$ for all k. Then, if $|x| \leq r_1 < t < r$,

$$| c_k x^k | \leq | c_k r_1{}^k | = | c_k t^k | \cdot \left| \frac{r_1}{t} \right|^k < A \left| \frac{r_1}{t} \right|^k .$$

But $r_1/t < 1$, so $\sum_{k=0}^{\infty} A | r_1/t |^k$ converges and the Weierstrass M-test applies. ∎

Before discussing continuity, differentiation, and integration of infinite series, let us review the tools at our disposal. [We shall use the following notation: f, f_k are functions in whatever space we are discussing; $s_n(x) = \sum_{k=1}^{n} f_k(x)$; $s(x)$ is the point wise limit of $s_n(x)$ as $n \to \infty$; $s_n \to s$ in the norm means $\lim_{n \to \infty} \| s_n - s \| = 0$, where $\|\cdot\|$ is the norm of the particular space.] By elementary calculus we know that if $f_k \in C[a, b]$, then $s_n \in C[a, b]$; if $f_k \in C^{(1)}[a, b]$, then $s_n \in C^{(1)}[a, b]$; and if f_k, $k = 1, 2, \ldots, n$, are Riemann integrable, s_n is Riemann integrable. We have proved the following:

(1) The space $C[a, b]$ is complete under the norm

$$\| f \| = \sup_{a \leq t \leq b} | f(t) | \qquad \text{(Theorem 4.1.2)}.$$

(2) The space $C^{(1)}[a, b]$ is complete under the norm

$$\| f \| = \sup_{a \leq t \leq b} | f(t) | + \sup_{a \leq t \leq b} | f'(x) | \qquad \text{(Theorem 5.2.2)}.$$

(3) If f_k, $k = 1, 2, \ldots, n$, is Riemann integrable on $[a, b]$, and $\lim_{n \to \infty} s_n(x) = s(x)$ *uniformly*, s is Riemann integrable and

$$\lim_{n \to \infty} \int_a^b s_n(x) \, dx = \int_a^b \lim_{n \to \infty} s_n(x) \, dx \qquad \text{(Theorem 6.4.4)}.$$

Thus, in order to discuss continuity, differentiability, or integrability of the limit function s, we need to know whether f_k is an element of whatever space is involved and whether $s_n \to s$ in the norm of that space ($s_n \to s$ uniformly in the case of integration).

7.1.4 EXAMPLE Find

$$\sum_{k=1}^{\infty} \int_0^{10} x(e^x - 1) e^{-kx} \, dx.$$

Solution   Let

$$s_n(x) = \sum_{k=1}^{n} x(e^x - 1)e^{-kx}$$

$$= x(e^x - 1)\,\frac{e^{-x}(1 - e^{-nx})}{(1 - e^{-x})}$$

(since $s_n(x)$ is a geometric series)

$$= x - xe^{-nx}.$$

So, $s(x) = x$, and we may conclude that

$$\sum_{k=1}^{\infty} \int_0^{10} f_k(x)\;dx = \int_0^{10} \sum_{k=1}^{\infty} f_k(x)\;dx = \int_0^{10} x\;dx$$

if $s_n \to s$ uniformly (that is, in the norm of $C[0, 10]$). But

$$\| s_n - s \| = \sup_{0 \le x \le 10} | -xe^{-nx} | = \sup_{0 \le x \le 10} xe^{-nx}.$$

The maximum value of xe^{-nx} occurs at $x = 1/n$ and is e^{-1}/n. So, $\| s_n - s \| = 1/en \to 0$ as $n \to \infty$. Therefore, $s_n \to s$ uniformly, and

$$\sum_{k=1}^{\infty} \int_0^{10} x(e^x - 1)e^{-kx}\;dx = \int_0^{10} x\;dx = 50.$$

7.1.5   *EXAMPLE*   If

$$f_k(x) = \frac{1}{k^3(1 + kx^3)}, \qquad 0 \le x \le 1,$$

show that

$$\frac{d}{dx}\left[\sum_{k=1}^{\infty} f_k(x) \right] = -3x^2 \sum_{k=1}^{\infty} \frac{1}{k^2(1 + kx^3)}.$$

Solution   Here

$$s_n(x) = \sum_{k=1}^{n} \frac{1}{k^3(1 + kx^3)}$$

and it is not convenient to find $s_n(x)$ in closed form. However,

$$\left| \frac{1}{k^3(1 + kx^3)} \right| \leq \frac{1}{k^3}$$

and $\sum_{k=1}^{\infty} 1/k^3$ converges, so $s_n \to s$ uniformly by the Weierstrass M-test. In order to conclude that $\| s_n - s \| \to 0$, where $\| \cdot \|$ is the norm in $C^{(1)}[0, 1]$, we must also show that $s_n' \to s'$ uniformly. But,

$$f_k'(x) = \frac{-3x^2}{k^2(1 + kx^3)}$$

so $|f_k'(x)| \leq 3/k^2$ and $\sum_{k=1}^{\infty} 3/k^2$ converges. Therefore,

$$\| s_n - s \| = \sup_{0 \leq x \leq 1} | s_n(x) - s(x) |$$

$$+ \sup_{0 \leq x \leq 1} | s_n'(x) - s'(x) | \to 0,$$

and since $C^{(1)}[0, 1]$ is complete, $s \in C^{(1)}[0, 1]$, and

$$\frac{d}{dx} \lim_{n \to \infty} s_n(x) = \lim_{n \to \infty} \frac{d}{dx} s_n(x).$$

PROBLEMS

1. Let $s_n(x) = \sum_{k=0}^{n} xe^{-kx}$ and $s(x) = \lim_{n \to \infty} s_n(x)$, $0 \leq x \leq 1$. Show that s is not the uniform limit of s_n by showing that s is not continuous.

2. Let $s_n(x) = \sum_{k=0}^{n} x^k(1 - x)$ and $s(x) = \lim_{n \to \infty} s_n(x)$. Determine whether s_n approaches s uniformly.

In Problems 3–5, justify the equalities.

3. $\displaystyle \int_1^2 \sum_{n=1}^{\infty} \frac{\ln nx}{n^2} \, dx = \sum_{n=1}^{\infty} \frac{\ln 4n - 1}{n^2}.$

4. $\displaystyle \int_0^{\pi} \sum_{n=1}^{\infty} \frac{n \sin nx}{e^n} \, dx = \frac{2e}{e^2 - 1}.$

5. $\dfrac{d}{dx}\left[\displaystyle\sum_{n=0}^{\infty} e^{-nx}\cos nx\right] = -\displaystyle\sum_{n=0}^{\infty} ne^{-nx}(\cos nx + \sin nx)$

for every $x > 0$.

In Problems 6–10, let $f_n(x)$ be such that the partial sums $s_n(x)$ are as given, $0 \leq x \leq 1$. Find $s(x) = \lim_{n\to\infty} s_n(x)$ (the pointwise limit of s_n). Determine whether this convergence is also norm convergence in (a) the space $C[0, 1]$, and (b) the space $C^{(1)}[0, 1]$.

6. $s_n(x) = \dfrac{x^n}{1 + x^n}.$

7. $s_n(x) = \dfrac{1}{n}\, e^{-nx} + x.$

8. $s_n(x) = x(1 - x)^n.$

9. $s_n(x)$
$= n(1 - x)e^{n(x-1)} + 1.$

10. $s_n(x) = nxe^{-nx} + \dfrac{e^{n(x-1)}}{n}.$

7.2 power series

An infinite series of the form $\sum_{n=0}^{\infty} c_n x^n$ or $\sum_{n=0}^{\infty} c_n(x - a)^n$ is called a *power series*. Although power series assume their most important role in the study of functions of complex variables, several results that we have at hand have useful applications to power series of real variables. The following example shows that elementary operations on polynomials may give rise to power series where the domain of convergence is more easily found than it is when more sophisticated methods are used.

7.2.1   EXAMPLE   By long division we have, for $t \neq -1$,

$$\frac{1}{1 + t} = 1 - t + t^2 - t^3 + \cdots + (-1)^{n-1}t^{n-1} + (-1)^n \frac{t^n}{t + 1}.$$

Integrating both sides from 0 to x,

$$\ln(1 + x) = x - \frac{x^2}{2} + \frac{x^3}{3} - \cdots + (-1)^{n-1}\frac{x^n}{n} + R_n,$$

$$\text{where}\quad R_n = (-1)^n \int_0^x \frac{t^n}{t+1}\, dt.$$

Now, if $0 \le x \le 1$,

$$|R_n| \le \int_0^x t^n\, dt = \frac{x^{n+1}}{n+1} \to 0 \qquad \text{as}\quad n \to \infty.$$

If $-1 < x < 0$, let $u = -t$. Then,

$$|R_n| \le \int_0^{-x} \frac{u^n}{1-u}\, du \le \frac{1}{1+x} \int_0^{-x} u^n\, du$$

$$= \frac{1}{1+x} \frac{|x|^{n+1}}{n+1} \to 0 \qquad \text{as}\quad n \to \infty.$$

Therefore,

$$\ln(1+x) = x - \frac{x^2}{2} + \frac{x^3}{3} - \cdots + (-1)^{n-1}\frac{x^n}{n} + \cdots$$

$$\text{for}\quad -1 < x \le 1.$$

The same expansion may be found by Maclaurin's formula (see Problem 8, Section 5.4), but both the Lagrange and Cauchy forms of the remainder are necessary in order to obtain the convergence set found here.

Considering the series for $1/(1+x)$ and $\ln(1+x)$, we see that integrating a series term-by-term may gain an end point in the convergence set, and, of course, differentiating may lose one. This situation at end points is one reason why we like to define the integral over a closed interval and the derivative over an open interval. Recall, however, that the radius of convergence of a power series (Section 2.5) was defined without regard to convergence or divergence at the end points of the interval of convergence. The following theorem assures us that differentiating and integrating a power series term-by-term may disturb convergence or divergence *only* at the end points of the interval of convergence:

7.2.2 THEOREM The series $\sum_{k=0}^{\infty} c_k x^k$ and $\sum_{k=1}^{\infty} k c_k x^{k-1}$ both have the same radius of convergence.

Proof   The proof is left as Problem 1.   ∎

Thus Theorem 7.1.3 applies not only to a given power series but also to its derived series, a fact that is essential for norm convergence in $C^{(1)}[a, b]$. We then know that *inside* its interval of convergence a power series may be differentiated or integrated term-by-term and the resulting series converges to the derivative or integral, respectively, of the function to which the original power series converges.

The following example suggests that Theorem 7.1.3 may be improved somewhat:

7.2.3   EXAMPLE   The geometric series $1 - t^2 + t^4 - t^6 + \cdots$ converges to $1/(1 + t^2)$ for $|t| < 1$, so the convergence is uniform for $|t| \leq r_1 < 1$. Therefore,

$$\tan^{-1} x = \int_0^x \frac{dt}{1 + t^2} = x - \frac{x^3}{3} + \frac{x^5}{5} - \frac{x^7}{7} + \cdots$$

for $0 \leq x < 1$.

But, the series $x - \frac{1}{3}x^3 + \frac{1}{5}x^5 - \cdots$ converges at $x = 1$, and the question arises: Does $\frac{1}{4}\pi = 1 - \frac{1}{3} + \frac{1}{5} - \cdots$? It can be shown (see Problem 2) by direct computation that the answer is yes, but this has not been established by a general theorem.

We give the following theorem, due to Abel, without proof:

7.2.4   THEOREM   If the power series $\sum_{k=0}^{\infty} a_k x^k$ converges for $0 \leq x \leq r$, then it converges uniformly for $0 \leq x \leq r$. Further, $\sum_{k=0}^{\infty} a_k x^k = f(x)$ is continuous on the left at $x = r$.

Thus, in the above example, $1 - \frac{1}{3} + \frac{1}{5} - \cdots$ converges to $\lim_{x \to 1-} \tan^{-1} x = \pi/4$.

In Section 2.7 the operations of addition, subtraction, and multiplication of absolutely convergent series were discussed, and the Weierstrass M-test showed that these results may be applied to power series inside the interval of convergence. The only algebraic question then remaining is that of division of power series. As may be expected, the problem of keeping the denominator away from zero necessitates, for the following theorem, a rather tedious proof which we omit.

7.2.5 *THEOREM* Let $\sum_{k=0}^{\infty} a_k x^k$ and $\sum_{k=0}^{\infty} b_k x^k$ converge for $|x| < r$, and let $b_0 \neq 0$. Then, for sufficiently small x,

$$f(x) = \frac{a_0 + a_1 x + a_2 x^2 + \cdots}{b_0 + b_1 x + b_2 x^2 + \cdots} = c_0 + c_1 x + c_2 x^2 + \cdots$$

converges, where the c's are found by the long division algorithm.

Thus, under appropriate conditions, algebraic operations, differentiation, integration, and Taylor's formula (Theorem 5.4.1) may all be used to obtain convergent power series. The following uniqueness theorems allow us to choose whichever means is most convenient, and, if the conditions for convergence are observed, to obtain *the* series expansion for a given function:

7.2.6 *THEOREM* If $f(x) = \sum_{k=0}^{\infty} a_k x^k$ has a (possibly infinite) positive radius of convergence,

$$a_n = \frac{f^{(n)}(0)}{n!} \qquad \text{for} \quad n = 0, 1, 2, \ldots .$$

Proof Set $x = 0$ in the series for $f(x), f'(x), f''(x), \ldots,$ $f^{(n)}(x), \ldots .$ ∎

7.2.7 *THEOREM* If $\sum_{k=0}^{\infty} a_k x^k = \sum_{k=0}^{\infty} b_k x^k = f(x)$, $\ |x| < r$, then $a_n = b_n$, $n = 0, 1, 2, \ldots .$

Proof By the previous theorem $a_n = f^{(n)}(0)/n! = b_n$. ∎

7.2.8 *EXAMPLE* By the above theorems,

$$\frac{e^x - 1}{x} = \frac{\left(1 + x + \dfrac{x^2}{2!} + \cdots \right) - 1}{x} = 1 + \frac{x}{2!} + \frac{x^2}{3!} + \cdots$$

for $x \neq 0$. Differentiating both sides of the equation,

$$\frac{xe^x - (e^x - 1)}{x^2} = \frac{1}{2!} + \frac{2x}{3!} + \frac{3x^2}{4!} + \cdots$$

or, for $x = 1$,

$$1 = \sum_{n=1}^{\infty} \frac{n}{(n+1)!}.$$

PROBLEMS

1. Prove Theorem 7.2.2. [*Hint*: show by the extended form of the root test that the radius of convergence in both cases is $R = 1/(\lim \sup_{n \to \infty} \sqrt[n]{|c_n|})$.]

2. Using the method of Example 7.2.1, show that

$$\int_0^x \frac{dt}{1+t^2} = x - \frac{x^3}{3} + \frac{x^5}{5} - \cdots \quad \text{for} \quad -1 < x \leq 1.$$

3. Note that $x - \frac{1}{3}x^3 + \frac{1}{5}x^5 - \cdots$ converges uniformly for $0 \leq x \leq 1$ by Theorem 7.2.4 and yet

$$\frac{d}{dx}\left(x - \frac{x^3}{3} + \frac{x^5}{5} - \cdots\right) = 1 - x^2 + x^4 - \cdots$$

does not converge if $x = 1$. Does this contradict Theorems 7.1.3 and 7.2.2? Explain.

4. Differentiate the following series twice, term by term, and discuss convergence of the series, the first derived series, and the second derived series. How does the situation here differ from that of power series?

$$\frac{\sin x}{1} + \frac{\sin 2x}{2^2} + \frac{\sin 3x}{3^2} + \cdots + \frac{\sin nx}{n^2} + \cdots.$$

In Problems 5–12, find a power series in x for the given function and state the interval of convergence.

5. $\tan x$.

6. $\sec x$.

8. $\ln \dfrac{1-x}{1+x}$.

7. $\sin^{-1} x$.

9. $\ln(\cos x)$.

10. $\dfrac{\sin^2 x}{1 - \cos x}.$ **12.** $\dfrac{\sin 2x}{\sin x}.$

11. $\dfrac{x - \sin x}{x^3}.$

Problems 13–20 refer to the functions

$$S(x) = x - \frac{x^3}{3!} + \frac{x^5}{5!} - \frac{x^7}{7!} + \cdots,$$

$$C(x) = 1 - \frac{x^2}{2!} + \frac{x^4}{4!} - \frac{x^6}{6!} + \cdots.$$

13. Prove that the above series converge absolutely for all x.
14. Prove $S'(x) = C(x)$ and $C'(x) = -S(x)$. Note the use of Theorem 7.2.2.
15. Show $S(0) = 0$, $C(0) = 1$. Find $S(x)C(y) + C(x)S(y)$, and show this equals $S(x + y)$.
16. Prove a property similar to that in Problem 15 for $C(x + y)$.
17. Prove $S^2(x) + C^2(x) = 1$ for all x.
18. Write

$$S(x) = x\left(1 - \frac{x^2}{2 \cdot 3}\right) + \frac{x^5}{5!}\left(1 - \frac{x^2}{6 \cdot 7}\right) + \cdots$$

and show $S(x) > 0$ for $0 < x < \sqrt{6}$. Similarly, show $C(x) > 0$ for $0 \le x < \sqrt{2}$.

19. Show by the properties of continuous functions that there is one and only one value ξ between $\sqrt{2}$ and 2 such that $C(\xi) = 0$. Let ξ be written $\pi/2$.

20. Show $S^2(\pi/2) = 1$, $S(\pi/4) = 1/\sqrt{2}$, $C(\pi/4) = 1/\sqrt{2}$.

21. Assuming the familiar properties of e^x except the series expansion, let

$$E(x) = 1 + x + \frac{x^2}{2!} + \frac{x^3}{3!} + \cdots.$$

Let $f(x) = E(x)/e^x$. Justify differentiating term-by-term and then show:

(a) $f'(x) = 0$,
(b) $f(x) = K$ (constant),
(c) $K = 1$.

VIII

LEBESGUE MEASURE

8.1 the measure of a bounded open set

We noted in Section 7.1 that the completeness of $C[a, b]$ and $C^{(1)}[a, b]$ justified the conclusion that the limit function of a sequence of continuous functions or continuously differentiable functions, respectively, was continuous or continuously differentiable, respectively, provided we had norm convergence in the norm of that particular space. Unfortunately, the situation in the case of Riemann integration was not so simple. If we required uniform convergence of a sequence of Riemann integrable functions, then the limit function was Riemann integrable. But uniform convergence is the norm convergence in $C[a, b]$ so we have done nothing more than add another operation to the functions in $C[a, b]$.

Many of the richest applications of integration involve discontinuous functions, and we have thus far been picking at the problem by noting (Theorem 6.1.7) that certain discontinuous functions are Riemann integrable and also (Problem 8, Section 6.4) that uniform convergence can be relaxed somewhat.

The only really comprehensive answer, however, lies in devising a different form of integration. We shall present the form due to Henri Lebesgue (1875–1941) whose name is also given to the kind of measure on which Lebesgue integration is based.

In order to motivate measure theory for its use in integration, recall from Chapter VI that for the existence of the Riemann integral we want values of the function to be close together for certain values of the independent variable. Thus, if $m_i = \inf |f(x)|$ and $M_i = \sup |f(x)|$ for $x_{i-1} \leq x \leq x_i$, we want $M_i - m_i$ to be small so that

$$\lim_{\Delta \to 0} | \overline{S_n} - \underline{s_n} | = \lim_{\Delta \to 0} \sum_{i=1}^{n} (M_i - m_i) \, \Delta x_i = 0,$$

where $\Delta x_i = x_i - x_{i-1}$ and $\Delta = \sup | x_i - x_{i-1} |$. This, roughly speaking, is the essence of continuous functions, so it is not surprising that a continuous function is Riemann integrable. We may ask if this state of affairs (keeping values of the function close together) can be accomplished in any way other than keeping values of the independent variable close together and requiring continuity.

The above question has no simple answer, but we shall use it as a point of departure in discussing measure theory. Thus, for a function f we are concerned with all values of x such that $A < f(x) < B$, where A and B are real numbers and $B - A$ is small. Such a set of values might be a collection of intervals and isolated points as pictured in Fig. 8, but it is not hard to imagine a more complicated set. Our problem is to devise a *measure* of such a set that can be used just as the *length* of an interval is used.

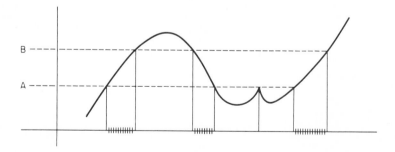

Figure 8

We begin with the measure of an interval, which should of course agree with our idea of length:

8.1.1   DEFINITION   If $I = (a, b)$, $(a, b]$, $[a, b)$ or $[a, b]$, the *measure* of I is $b - a$. We write $m(I) = b - a$.

Thus the measure is defined to be the same whether the end points are included or not. We shall generally use open intervals, but when the need arises for a closed interval (as, for example, if the Heine–Borel theorem is to be used), the measure will be the same as that of the corresponding open interval. A trivial consequence of the work in this chapter will be that $m([a, b]) = m((a, b)) + m(\{a\}) + m(\{b\}) = (b - a) + 0 + 0$ so no inconsistencies are introduced by this definition.

In order to extend the idea of the measure of an open interval to the measure of an open set, the following theorem about sets is fundamental.

8.1.2   THEOREM   Any bounded open set G of real numbers may be written as the union of a countable number of disjoint open intervals.

Proof   Let G be a bounded open set of real numbers contained in the interval $[a, b]$, and let $x \in G$. Then, by Definition 4.3.1, $(x, x + \delta) \subset G$ for sufficiently small, positive δ. Let δ_1 be the least upper bound of these δ's. Note that $x + \delta_1 \notin G$. (Why?) Similarly, find a point $x - \delta_2$ such that $(x - \delta_2, x) \subset G$, but $x - \delta_2 \notin G$. Thus $(x - \delta_2, x + \delta_1) \subset G$ so every point of G is in one and only one interval, and G is the union of these intervals. To see that the intervals are disjoint, suppose y is in $(x - \delta_2, x + \delta_1)$ and also in $(x - \delta_4, x - \delta_3)$ both formed as described. Then $x - \delta_4 < x + \delta_1$ and $x - \delta_2 < x + \delta_3$. Since $x - \delta_2 \notin G$, $x - \delta_2 \leq x - \delta_4$. Similarly, since $x - \delta_4 \notin G$, $x - \delta_4 \leq x - \delta_2$. Thus $x - \delta_4 = x - \delta_2$. Similarly, $x + \delta_3 = x + \delta_1$. To see that these are countable, first pick the interval, if there is one, whose length is greater than $\frac{1}{2}(b - a)$. Then pick those, from left to right, whose lengths are greater than $\frac{1}{3}(b - a)$ but less than or equal to $\frac{1}{2}(b - a)$. Then choose those whose lengths are greater than

$\frac{1}{4}(b - a)$ but less than or equal to $\frac{1}{3}(b - a)$. Continuing in this way, the intervals are put into a one-to-one correspondence with the positive integers. ▮

8.1.3 *DEFINITION* For a bounded open set $G = \bigcup_{k=1}^{\infty} I_k$, where each I_k is an open interval and $I_k \cap I_j = \emptyset$ if $j \neq k$, $m(G) = \sum_{k=1}^{\infty} m(I_k)$.

Note that the decomposition of G into the union of disjoint open intervals is guaranteed by the preceding theorem, and the infinite series $\sum_{k=1}^{\infty} m(I_k)$ converges since $m(I_k) \geq 0$ and

$$\sum_{k=1}^{\infty} m(I_k) \leq b - a.$$

(See Problem 5.)

PROBLEMS

1. Let $f(x) = x^2$. Describe the following sets of real numbers:

   (a) $S = \{x \mid 4 \leq f(x) \leq 9\}$,
   (b) $R = \{x \mid -1 < f(x) < +1\}$.

2. Let $f(x) = x^3 - 4x$. Describe the set S of real numbers if

   $$S = \{x \mid 0 \leq f(x) < 15\}.$$

3. Let $f(x) = x^3 - 3x$. Find the relative maximum and minimum values of $f(x)$, and then describe the set $S = \{x \mid 2 \leq f(x) < 18\}$.

4. Let

   $$f(x) = \begin{cases} 1 & \text{if } x \text{ is irrational,} \\ 0 & \text{if } x \text{ is rational.} \end{cases}$$

   Find:

   (a) $\left\{ x \mid \dfrac{1}{\sqrt{3}} \leq f(x) \leq \dfrac{1}{\sqrt{2}} \right\}$,

(b) $\left\{x \,\middle|\, \dfrac{1}{2} < f(x) < \dfrac{3}{2}\right\}$,

(c) $\left\{x \,\middle|\, -\dfrac{1}{2} < f(x) < +\dfrac{1}{2}\right\}$.

5.   Verify that $\sum_{k=1}^{\infty} m(I_k) \le b - a$ in Definition 8.1.3.

6.   Let $G = \bigcup_{k=1}^{\infty} I_k$, where $I_k = (1/(k+1),\, 1/k)$. Find $m(G)$.
     (*Hint*:   see Example 2.3.6.)

7.   The measure of a general set of points has not yet been defined;
     but, from Problem 6 what can be anticipated about the measure
     of the set $\{\frac{1}{2},\, \frac{1}{3},\, \frac{1}{4},\, \ldots\}$?

8.   Let $G = \bigcup_{k=1}^{\infty} I_k$, where $I_k = [1/(k+1)^2,\, 1/k^2]$. Find $m(G)$.

9.   Let $G = \bigcup_{k=1}^{\infty} I_k$, where $I_k = (3/2^{k+1},\, 1/2^{k-1})$. Find $m(G)$.

10. If $G = \bigcup_{k=2}^{\infty} I_k$, where $I_k = (1/k^2,\, 1/k)$, is G open? What can
     be said about $m(G)$?

8.2 outer measure—the measure of a bounded set

The analogy between measure and length, or between measure and
area (in two dimensions) is a very useful one but also a somewhat
dangerous one because of the very general meanings of length and
area in common usage. Thus, one speaks of "a vast area of undeveloped
land," where the words "space" or "expanse" might better be used
instead of "area." In our usage, "area" or "length" or "measure"
will mean the real number (or possibly $+\infty$) attached to the geometric configuration according to certain rules or definitions. Thus,
it is conceivable that a certain configuration might not have a
"length" or an "area" or a "measure" if it is not possible to attach
a number to that certain set of points according to the rules.

We have thus far defined the measure of an open set of real numbers
in terms that certainly exist. In order to define the measure of a
one-dimensional set of points in general, we shall first define the
outer measure and inner measure. It should be observed that these

always exist, although the measure, when it is defined, might fail to exist.

8.2.1 *DEFINITION*   Let E be a bounded set of points in the interval $[a, b]$. The outer measure of E is the greatest lower bound, or infimum, of the measures of the open sets containing E. We write

$$m^{\oplus}(E) = \inf_{G \supset E} m(G), \qquad G \quad \text{open.}$$

8.2.2 *DEFINITION*   Let E be a bounded set of points in the interval $[a, b]$. The inner measure of E is

$$m_{\oplus}(E) = b - a - m^{\oplus}(CE),$$

where CE is the complement of E with respect to $[a, b]$.

8.2.3 *DEFINITION*   If $m^{\oplus}(E) = m_{\oplus}(E)$, E is measurable, and we write
$$m(E) = m^{\oplus}(E) = m_{\oplus}(E).$$

We collect some elementary facts about the measure of a bounded set in the following theorem.

8.2.4 *THEOREM*   For any bounded set E in the interval $[a, b]$ and CE the complement of E with respect to $[a, b]$, we have:

(a) $0 \leq m^{\oplus}(E) \leq b - a$;

(b) $m_{\oplus}(CE) = b - a - m^{\oplus}(E)$;

(c) If E is measurable, CE is measurable and $m(E) + m(CE) = b - a$;

(d) If E is open, $m^{\oplus}(E)$ is the same as $m(E)$ given in Definition 8.1.3.

Proof   Part (a) follows directly from the definitions; (b) and (c) depend only on observing that $C(CE) = E$; (d) is left as Problem 1.   ∎

It should be noted that we have given two (possibly different) definitions that can be used for the measure of an open set (Definitions 8.1.3 and 8.2.3). Part (d) in the above theorem goes only part way in proving that they are equivalent, so we should probably use different notations for the two different definitions. However, since we can complete the proof of their equivalence in the next section, we ask the student to note that in the meantime only the constructive definition (Definition 8.1.3) will be used when referring to open sets.

The following theorem seems obvious from the definition, or at least it should be anticipated from the terminology, but some proof is needed.

8.2.5   *THEOREM*   If E is any bounded set in $[a, b]$, then

$$m_\oplus(E) \leq m^\oplus(E).$$

Proof   From the definition of $m^\oplus$, there are open sets G and G' containing E and CE, respectively, such that

$$m(G) < m^\oplus(E) + \epsilon \qquad \text{and} \qquad m(G') < m^\oplus(CE) + \epsilon,$$

where ϵ is any positive number. Now consider the closed interval $[a + \epsilon, b - \epsilon]$. We know that

$$G = \bigcup_{k=1}^{\infty} I_k, \qquad I_k \cap I_j = \emptyset, \quad \text{if} \quad j \neq k,$$

and

$$G' = \bigcup_{k=1}^{\infty} I_k', \qquad I_k' \cap I_j' = \emptyset, \quad \text{if} \quad j \neq k.$$

Then $\{I_k\}$ and $\{I_k'\}$ together cover $[a + \epsilon, b - \epsilon]$ so, by the Heine–Borel theorem a finite number of these intervals will cover $[a + \epsilon, b - \epsilon]$. Let H be the union of this finite number of intervals. Then $m(H) \geq b - a - 2\epsilon$ and $m(H) \leq m(G) + m(G')$ (see Problem 4). Then,

$$m^\oplus(E) + m^\oplus(CE) + 2\epsilon \geq m(G) + m(G') \geq b - a - 2\epsilon$$

or

$$m^\oplus(E) + m^\oplus(CE) \geq b - a - 4\epsilon.$$

But ϵ was arbitrary, so $m^{\oplus}(E) + m^{\oplus}(CE) \geq b - a$. Thus

$$m^{\oplus}(E) \geq b - a - m^{\oplus}(CE) = m_{\oplus}(E). \quad \blacksquare$$

8.2.6 *REMARK*   Since $m^{\oplus}(E)$ and $m^{\oplus}(CE)$ lie between 0 and $b - a$, inclusive, and since $m_{\oplus}(E) = b - a - m^{\oplus}(CE) \leq m^{\oplus}(E)$, $m_{\oplus}(E)$ also lies between 0 and $b - a$, inclusive. Therefore, if $m^{\oplus}(E) = 0$, we also have $m_{\oplus}(E) = 0$, so E is measurable and $m(E) = 0$.

8.2.7 *THEOREM*   Let E be a countable, bounded set. Then E is measurable and $m(E) = 0$.

Proof   Let $E = \{x_1, x_2, \ldots\}$, and fix $\epsilon > 0$. There is an open interval of length $\epsilon/2$ which contains x_1; there is an open interval of length $\epsilon/4$ which contains x_2; and, in general, there is an open interval of length $\epsilon/2^n$ which contains x_n. Thus E is contained in an open set whose measure is less than or equal to ϵ. But, ϵ was arbitrary, so $m^{\oplus}(E) = 0$. By the above remark, $m(E) = 0$. $\blacksquare$

Note: the above theorem is true for E unbounded, also, with no modification in the proof, as will be seen following Section 8.3.

PROBLEMS

1. Let E be a bounded open set of real numbers. Show that Definitions 8.1.3 and 8.2.1 give the same result for E. What is necessary in order to conclude that Definitions 8.1.3 and 8.2.3 are equivalent?

2. Let E_1 and E_2 be two sets with $E_1 \cap E_2 = \emptyset$. Prove $m(E_1 \cup E_2) = m(E_1) + m(E_2)$ if

    (a) E_1 and E_2 are bounded open intervals,
    (b) E_1 and E_2 are bounded open sets.

3. Let I_1 and I_2 be two bounded open intervals. Prove $m(I_1 \cup I_2) = m(I_1) + m(I_2) - m(I_1 \cap I_2)$.

4. In the proof of Theorem 8.2.5, supply the details to show that $m(H) \geq b - a - 2\epsilon$ and $m(H) \leq m(G) + m(G')$. Why was the Heine–Borel theorem used in the proof of this theorem instead of leaving an infinite number of intervals with which to deal?

5. Let $A \subset B$. Show that $m^{\oplus}(A) \leq m^{\oplus}(B)$. State and prove a relationship between $m_{\oplus}(A)$ and $m_{\oplus}(B)$.

6. Let A be a measurable set. Define

$$A + c = Q = \{x + c \mid x \in A\} = \{x \mid (x - c) \in A\}.$$

Then Q is called the *translation of A by c*. Prove Q is measurable.

Problems 7–10 refer to a circle Q of radius 1 inscribed in a square S as in Fig. 9. Let the area of polygons in two dimensions take the place of the measure of open sets in Definitions 8.2.1 and 8.2.2. Also, the area of S replaces the length of the interval $[a, b]$.

7. Show that the area of a regular polygon of n sides circumscribed about Q is $n \tan \pi/n$.

8. Show that the area of a regular polygon of n sides inscribed in Q is $(n/2) \sin(2\pi/n)$.

9. Verify from the definitions and above instructions that

$$m^{\oplus}(Q) = \inf\left(n \tan \frac{\pi}{n}\right)$$

and

$$m_{\oplus}(Q) = 4 - \left[\inf\left(4 - \frac{n}{2} \sin \frac{2\pi}{n}\right)\right] = \sup\left(\frac{n}{2} \sin \frac{2\pi}{n}\right),$$

where both inf and sup occur as $n \to \infty$.

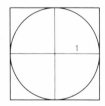

Figure 9

10.   Show $m^{\oplus}(Q) = m_{\oplus}(Q) = \pi$ and thus Q is measurable.

11.   Let $H = \bigcup_{k=1}^{\infty} I_k$, where $I_k = [1/(k+1)^2, 1/k^2]$. From the definitions, find $m^{\oplus}(H)$ and $m_{\oplus}(H)$. (See Problem 8, Section 8.1.) To find $m_{\oplus}(H)$, let $[a, b]$ be $[0, 1]$.

12.   Let $H = \bigcup_{k=1}^{\infty} I_k$, where $I_k = [3/2^{k+1}, 1/2^{k-1}]$. From the Definitions find $m^{\oplus}(H)$ and $m_{\oplus}(H)$. (See Problem 9, Section 8.1.) To find $m_{\oplus}(H)$, let $[a, b]$ be $[0, 1]$.

13.   In Problems 11 and 12 does $m_{\oplus}(H)$ depend on the interval $[a, b]$? Justify your answer in the general case.

8.3 the class of measurable sets

In this section we shall investigate how large a class of measurable sets we have created and, as a by-product, show that an open set is measurable according to Definition 8.2.3. Then, since the outer measure of an open set agrees with the constructive definition of measure in 8.1.3, we have no further worries about two (possibly different) definitions of measure applied to open sets.

It turns out, not unexpectedly, that we can combine measurable sets by a finite or a countably infinite number of unions, intersections, or complements and still have a measurable set.

Before considering this statement in detail, let us digress in order to take care of two minor points of notation and terminology. First, it has probably been apparent that the word *complement* may be ambiguous unless it is stated or clearly implied what set or space acts as the universe with respect to which we take the complement. Thus, if A is a set of real numbers, CA may be the complement of A with respect to the space R of all real numbers; or, if $A \subset I$, an interval, the complement may be taken with respect to I. In order to make this clear, the following notation is useful.

8.3.1   *NOTATION*   If $A \subset Q$, $Q - A$ is the set of real numbers in Q but not in A. If $A \not\subset Q$, $Q - A = Q - (A \cap Q)$. If no ambiguity can arise, $Q - A$ may be written CA.

Second, all sets of numbers considered thus far have been bounded, that is they are contained in an interval $[a, b]$ or (a, b). This restriction has assured the convergence of infinite sums like $\sum_{k=1}^{\infty} m(I_k) = m(G)$ in Definition 8.1.3 but has prevented us from considering sets like the following: Let the integer 1 be surrounded by an open interval of length $\epsilon/2$; let 2 be surrounded by an open interval of length $\epsilon/4$; and, in general, let n be surrounded by an interval of length $\epsilon/2^n$. Call the union of these open intervals S. Then S is open by Theorem 4.3.6, and we can see its decomposition into open intervals. Even though S is unbounded, it seems reasonable to have $m(S) = \epsilon(\frac{1}{2} + \frac{1}{4} + \frac{1}{8} + \cdots) = \epsilon$.

The above example can be considered by means of a limiting process, to which we return at the end of this section. Briefly, we will want all bounded subsets of an unbounded set to be measurable, but if that condition is satisfied we will allow any nonnegative number or $+\infty$ for the measure of unbounded sets. As usual, care will have to be taken to avoid meaningless expressions like $\infty - \infty$.

The following theorem gives the fundamental results concerning the class of measurable sets of real numbers. Its proof involves a sequence of results that are given in lemmas following the statement of the theorem.

8.3.2 *THEOREM* If E_1, E_2, ... are bounded measurable sets, and $E = \bigcup_{i=1}^{\infty} E_i$ is bounded, then

(a) E is measurable and $m(E) \leq \sum_{i=1}^{\infty} m(E_i)$,

(b) $\bigcap_{i=1}^{\infty} E_i$ is measurable.

8.3.3 *LEMMA* If G_1, G_2, ..., and $G = \bigcup_{i=1}^{\infty} G_i$ are bounded open sets, then

$$m(G) \leq \sum_{i=1}^{\infty} m(G_i).$$

Proof By Theorem 8.1.2, each of the open sets may be written as the union of a countable number of disjoint open intervals. Let

$$G_n = \bigcup_{m=1}^{\infty} I_{mn}, \quad \text{where} \quad I_{mn} = (a_{mn}, b_{mn}), \quad \begin{matrix} n = 1, 2, \ldots, \\ m = 1, 2, \ldots, \end{matrix}$$

and $I_{mn} \cap I_{m'n} = \emptyset$ if $m \neq m'$; and let

$$G = \bigcup_{k=1}^{\infty} I_k, \quad \text{where} \quad I_k = (\alpha_k, \beta_k), \quad I_k \cap I_{k'} = \emptyset \text{ if } k \neq k'.$$

Since $G = \bigcup_{n=1}^{\infty} G_n$, $[\alpha_k + \epsilon, \beta_k - \epsilon]$ is covered by the open intervals $\{I_{mn}\}$, $m = 1, 2, \ldots, n = 1, 2, \ldots$. By the Heine–Borel theorem, a finite number of these intervals will cover $[\alpha_k + \epsilon, \beta_k - \epsilon]$ for each k. Let H be the union of these intervals so that, as in Theorem 8.2.5, $\beta_k - \alpha_k - 2\epsilon \leq m(H)$. Now, $H = \bigcup_{j=1}^{p} H_j$, where each H_j is the same as some I_{mn}, and $m(H) \leq \sum_{j=1}^{p} m(H_j)$. Then,

$$\beta_k - \alpha_k \leq \sum_{j=1}^{p} m(H_j) + 2\epsilon,$$

so

$$m(G) = \sum_{k=1}^{\infty} (\beta_k - \alpha_k) \leq \sum_{k=1}^{\infty} \left[\sum_{j=1}^{p} m(H_j) \right] + 2\epsilon.$$

Now, the above series are absolutely convergent so we may pick the following rearrangement (with possibly some additional terms included which will not affect the inequality).

$$\sum_{n=1}^{\infty} \left[\sum_{m=1}^{\infty} (b_{mn} - a_{mn}) \right].$$

Then, since ϵ is arbitrary, $m(G) \leq \sum_{n=1}^{\infty} m(G_n)$. ∎

8.3.4 *COROLLARY* If in the above lemma, $G_i \cap G_j = \emptyset$, $i \neq j$, then

$$m(G) = \sum_{i=1}^{\infty} m(G_i).$$

Proof In the proof of Lemma 8.3.3, each (α_k, β_k) coincides exactly with some (a_{mn}, b_{mn}) so the inequalities become equalities. ∎

8.3.5 *LEMMA* If $E_1, E_2, \ldots$, and $E = \bigcup_{i=1}^{\infty} E_i$ are bounded sets, then

$$m^{\oplus}(E) \leq \sum_{i=1}^{\infty} m^{\oplus}(E_i).$$

Proof By the definition of m, we can put E_n in an open set G_n such that

$$m(G_n) < m^{\oplus}(E_n) + \frac{\epsilon}{2^n}.$$

Let $G = \bigcup_{n=1}^{\infty} G_n$. Then,

$$m(G) \leq \sum_{n=1}^{\infty} m(G_n) < \sum_{n=1}^{\infty} m^{\oplus}(E_n) + \epsilon.$$

But, G is open and $G \supset E$. Therefore, $m^{\oplus}(E) \leq m(G)$. Thus, $m^{\oplus}(E) < \sum_{n=1}^{\infty} m^{\oplus}(E_n) + \epsilon$. But ϵ was arbitrary, so the result follows. ∎

8.3.6 *LEMMA* If $E_1, E_2, \ldots$ are bounded measurable sets and if $E = \bigcup_{i=1}^{\infty} E_i$ is bounded and $E_i \cap E_j = \emptyset$ if $i \neq j$, then E is measurable and

$$m(E) = \sum_{i=1}^{\infty} m(E_i).$$

Proof We prove this for two sets and leave the rest for the problems. Let $E = E_1 \cup E_2$, where E_1 and E_2 are measurable sets. We know from the previous lemma that

$$m^{\oplus}(E) \leq m^{\oplus}(E_1) + m^{\oplus}(E_2) = m(E_1) + m(E_2).$$

We shall now prove that $m_{\oplus}(E) \geq m(E_1) + m(E_2)$ and the result will follow. Let all sets be contained in the interval $[a, b]$. Put CE_1 and CE_2 in open sets G_1 and G_2, respectively, such that

$$m(G_1) < m(CE_1) + \frac{\epsilon}{2} \quad \text{and} \quad m(G_2) < m(CE_2) + \frac{\epsilon}{2}.$$

Since $E_1 \cap E_2 = \emptyset$, $CE_1 \cup CE_2 = [a, b]$ and so also $G_1 \cup G_2 \supset (a, b)$. Now,

$$m(G_1 \cap G_2) \le m(G_1) + m(G_2) - (b - a)$$

(see Problem 1). But, $CE \subset G_1 \cap G_2$. Therefore,

$$m^{\oplus}(CE) \le m(G_1 \cap G_2) \le m(G_1) + m(G_2) - (b - a)$$
$$< m(CE_1) + m(CE_2) + \epsilon$$
$$- (b - a).$$

But, ϵ was arbitrary, so

$$m^{\oplus}(CE) \le m(CE_1) + m(CE_2) - (b - a)$$

or

$$m_{\oplus}(E) = b - a - m^{\oplus}(CE) \ge (b - a) - m(CE_1)$$
$$+ (b - a) - m(CE_2)$$
$$= m(E_1) + m(E_2),$$

which completes the proof. ∎

If, in the above theorem, E_1, E_2, ... are taken to be the open intervals forming a general open set, we see that an open set is measurable according to Definition 8.2.3, and thus, as noted at the beginning of this section, the two definitions agree for open sets.

8.3.7 *COROLLARY*   If E_1 and E_2 are bounded measurable sets and $E_1 \subset E_2$, then $E_2 - E_1$ is measurable.

Proof   Write $C(E_2 - E_1) = E_1 \cup CE_2$. Then, since complements and unions of measurable sets are measurable, $E_2 - E_1$ is measurable. ∎

8.3.8 *LEMMA*   If E and H are bounded measurable sets, then $E \cap H$ is measurable.

Proof   Let $E \subset [a, b]$, $H \subset [a, b]$, and suppose first that H is an open interval (α, β). Let $E_1 = E \cap H$, $E_2 = E - E_1$, so $E = E_1 \cup E_2$, $E_1 \cap E_2 = \emptyset$. Let G be open, $G \supset E$, $G_1 = G \cap H$, $G_2 = (G - G_1) - Q$, where Q consists of the singleton sets $\{\alpha\}$ or $\{\beta\}$ or both—whatever is necessary to make G_2 open. In any case, $m(Q) = 0$, and $G = G_1 \cup G_2 \cup Q$, so $m(G) = m(G_1) + m(G_2)$. Taking the greatest lower bounds of such measures with the G's described as above we have $m^\oplus(E) = m^\oplus(E_1) + m^\oplus(E_2)$. Similarly, let

$$E' = CE, \quad E_1' = E' \cap H, \quad E_2' = E' - E_1', \quad E' = E_1' \cup E_2'.$$

As before, $m^\oplus(E') = m^\oplus(E_1') + m^\oplus(E_2')$. Now, E was assumed to be measurable, so $m^\oplus(E) + m^\oplus(E') = b - a$, and

$$m^\oplus(E_2) + m^\oplus(E_2') \geq m^\oplus(E_2 \cup E_2') = (b - a) - (\beta - \alpha).$$

Therefore, $m^\oplus(E_1) + m^\oplus(E_1') \leq \beta - \alpha$. Moreover,

$$\beta - \alpha = m(H) = m^\oplus(E_1 + E_1') \leq m^\oplus(E_1) + m^\oplus(E_1').$$

But, $m_\oplus(E_1) = \beta - \alpha - m^\oplus(E_1')$, so $m_\oplus(E_1) = m^\oplus(E_1)$.

This proves the theorem if H is an open interval, and the proof can obviously be extended to the case of an open set by the previous lemma. In the general case, let H be included in an open set G and CH in an open set G', so that $m(G) + m(G') < b - a > \epsilon$. It can then be shown (Problem 3) that

$$m^\oplus(E \cap H) + m^\oplus(C(E \cap H)) \leq b - a$$

so that $E \cap H$ is measurable.   ∎

8.3.9   COROLLARY   If E_1 and E_2 are measurable, $E_1 - E_2$ is measurable.

Proof   The proof is left as Problem 4.   ∎

8.3.10   PROOF OF THEOREM 8.3.2

(a) Let $E = \bigcup_{i=1}^{\infty} E_i$, $E_1' = E_1$, $E_2' = E_2 - E_1$, $E_3' = E_3 - (E_1 \cup E_2)$, ..., $E_n' = E_n - (\bigcup_{i=1}^{n-1} E_i)$. Now, E_n' $(n = 1, 2, \ldots)$ is measurable by the above lemmas. Also, $E = \bigcup_{i=1}^{\infty} E_i'$, and $E_i' \cap E_j' = \emptyset$, $i \neq j$. So, E is measurable. The inequality follows from Lemma 8.3.5.

(b) If $H = \bigcap_{i=1}^{\infty} E_i$, $CH = \bigcup_{i=1}^{\infty} CE_i$, so CH is measurable, and therefore H is measurable. ∎

It might be asked at this point whether all sets are measurable, and the answer is no. That is, it is possible to have a set whose inner and outer measures are not the same. Unfortunately, in order to demonstrate such a set, it is necessary to appeal to the Axiom of Choice. A full discussion of this axiom, its consistency with other axioms in set theory, and its implications is an interesting digression which we shall not pursue here. The student is referred, for example, to the work of Halmos. [†]

Briefly, the Axiom of Choice allows us to form an infinite set each of whose elements comes from exactly one set in a nondenumerably infinite collection of sets. If this collection were finite, or denumerable, there would be no problem, because we could appeal to mathematical induction; so in a sense the Axiom of Choice extends our familiar reliance on induction to the nondenumerable case.

Let us return to the construction of a nonmeasurable set. For simplicity we shall construct it on the circumference of a circle and then "unroll" the circle to fit the interval $[0, \pi]$.

Let the diameter of the circle be 1. Divide the circumference of the circle into sets with the property that if P is any point in a given set the points that are a distance $1, 2, \ldots$ in either direction from P are in the same set. Note that (1) the sets contain an infinite number of points since the circumference π is not rational, (2) there are a nondenumerably infinite number of sets, (3) two sets are either disjoint or identical, and (4) their union is the whole circumference. We shall consider identical sets only once.

Now, take one representative point from each set, by the Axiom of Choice, and call this new set of points E. Let E_n, E_{-n} be the sets obtained from E by n rotations in the clockwise and counterclockwise directions. Then E, E_1, E_{-1}, E_2, E_{-2}, $\ldots$ are disjoint and together fill the circumference of the circle.

When these are "unrolled" on the straight line, they fill the interval $[0, \pi]$, they are congruent with each other, and they must therefore have the same measure. But we now have a countable number of

[†] Halmos, P. R., "Naive Set Theory." Van Nostrand-Reinhold, Princeton, New Jersey, 1960.

disjoint sets to measure, so the measure of the union is the sum of the measures of the sets. But the measure of the union of the sets is π.

Thus we have arrived at a contradiction, for if the measure of each set is zero, the measure of their union is zero. On the other hand, if the measure of each set is not zero, it must be some positive number so the measure of their union is ∞. Since neither case is possible, there exists a nonmeasurable set.

Let us return now to the problem of defining the measure of an unbounded set of real numbers.

8.3.11   DEFINITION   Let E be an unbounded set of real numbers, and let I_n be the interval $[-n, n]$ for $n = 1, 2, \ldots$. If the sets $E \cap I_n$ are measurable for $n = 1, 2, \ldots$, then E is said to be *measurable* and

$$m(E) = \lim_{n \to \infty} [m(E \cap I_n)].$$

In the above definition note that $m(E \cap I_n)$ is a monotone nondecreasing sequence of real numbers, so if $+\infty$ is allowed as a measure of certain sets, $m(E)$ always exists provided $(E \cap I_n)$ is measurable for each n.

It is now easy to verify that every countable set has measure zero whether bounded or not. If E is countable, we know from Theorem 8.2.7 that $m(E \cap I_n) = 0$ for every n, so $\lim_{n \to \infty}(E \cap I_n) = 0$.

We also wish to verify that the major result of this section holds for unbounded sets.

8.3.12   THEOREM   Parts (a) and (b) of Theorem 8.3.2 hold for unbounded measurable sets.

Proof   If $E_1, E_2, \ldots$, and $E = \bigcup_{i=1}^{\infty} E_i$ are measurable, then $(E_1 \cap I_n), (E_2 \cap I_n), \ldots$, and $(E \cap I_n)$ are all measurable, for $n = 1, 2, \ldots$.

(a) For $n = 1, 2, \ldots$,

$$(E \cap I_n) = \left(\left[\bigcup_{i=1}^{\infty} E_i \right] \cap I_n \right) = \left(\bigcup_{i=1}^{\infty} [E_i \cap I_n] \right),$$

so by Theorem 8.3.2

$$m(E \cap I_n) = m\left(\bigcup_{i=1}^{\infty} [E_i \cap I_n]\right) \le \sum_{i=1}^{\infty} m(E_i \cap I_n)$$

for $n = 1, 2, \ldots$.

Taking limits as $n \to \infty$,

$$m(E) = \lim_{n \to \infty} m(E \cap I_n)$$

$$\le \lim_{n \to \infty} \left[\sum_{i=1}^{\infty} m\,(E_i \cap I_n) \right]$$

$$= \lim_{n \to \infty} \left[\lim_{k \to \infty} \left\{ \sum_{i=1}^{k} m(E_i \cap I_n) \right\} \right]$$

$$= \lim_{k \to \infty} \left[\lim_{n \to \infty} \left\{ \sum_{i=1}^{k} m(E_i \cap I_n) \right\} \right] \qquad \text{(see Problem 6,}$$
$$\text{Section 3.3)}$$

$$= \lim_{k \to \infty} \left[\sum_{i=1}^{k} \{ \lim_{n \to \infty} m(E_i \cap I_n) \} \right]$$

$$= \sum_{i=1}^{\infty} m(E_i).$$

(b) Note that

$$\left(\bigcap_{i=1}^{\infty} E_i\right) \cap I_n = \bigcap_{i=1}^{\infty} (E_i \cap I_n) \qquad \text{for} \quad n = 1, 2, \ldots,$$

and the result follows from Theorem 8.3.2 (b). ∎

8.3.13 *EXAMPLE* Let U be the nonmeasurable set formed above in the interval $[0, \pi]$, and let $V = (\pi, \infty)$. Then $m(V) = \infty$. However, $S = U \cup V$ is not measurable because the intersection of S with $[0, n]$ is not measurable.

It should be noted that the clue to measurability of a general set E is the measurability of intersections of E with sets whose measure properties are well known. This is the motivation for a development of measure theory, due to Carathéodory, which is perhaps more elegant but less intuitive than that just presented. Recall that $m^{\oplus}(A)$ exists for every bounded A. This is easily extended, by the same definition as given for the bounded case, to unbounded sets. So, $m^{\oplus}(A)$ exists for all A. Then, a set E is said to be *measurable* if for each set A we have $m^{\oplus}(A) = m^{\oplus}(A \cap E) + m^{\oplus}(A \cap CE)$; and then let $m(E) = m^{\oplus}(E)$. Then, theorems similar to those presented here can be proved with the new definition. Obviously these will apply to the unbounded case without any additional work.

PROBLEMS

1. Let $I = [a, b]$, $G_1 \subset I$, $G_2 \subset I$, $I \subset G_1 \cup G_2$. Supply the proof, needed in Lemma 8.3.6, that

$$m(G_1 \cap G_2) \le m(G_1) + m(G_2) - (b - a).$$

2. Complete the proof of Lemma 8.3.6.

3. In Lemma 8.3.8, complete the proof that

$$m^{\oplus}(E \cap H) + m^{\oplus}(C(E \cap H)) \le b - a.$$

4. Prove Corollary 8.3.9.

5. If E_1 and E_2 are measurable sets, show
$$m(E_1 \cup E_2) = m(E_1) + m(E_2) - m(E_1 \cap E_2).$$

6. State and prove a proposition for three measurable sets similar to Problem 5.

7. For two sets A and B, define $A \triangle B = (A - B) \cup (B - A)$. Prove that if A and B are measurable, $A \triangle B$ is measurable.

8. A σ-ring is a nonempty class S of sets such that

   (a) if $E \in S$ and $F \in S$, then $E - F \in S$, and
   (b) if $E_i \in S$, $i = 1, 2, \ldots$, then $\bigcup_{i=1}^{\infty} E_i \in S$.

   Prove that the class of measurable sets is a σ-ring.

9. Let S be the smallest σ-ring containing all open intervals in $[a, b]$. The members of S are called *Borel sets*. Are all Borel sets measurable? Are all measurable sets Borel sets?

10. Let $E \subset G = \bigcap_{i=1}^{\infty} G_i$, where G_i $(i = 1, 2, \ldots)$ is open. If $m^{\oplus}(G - E) = 0$, prove E is measurable.

11. Let $E \supset F = \bigcup_{i=1}^{\infty} F_i$, where F_i $(i = 1, 2, \ldots)$ is closed. If $m^{\oplus}(E - F) = 0$, prove E is measurable.

12. Let $m(A) = 0$, and let B be a Borel set. Prove $A \bigtriangleup B$ is measurable. Is A necessarily a Borel set?

13. The Cantor set C is formed as follows: Remove the open middle third from the interval $[0, 1]$, then the open middle third from $[0, \frac{1}{3}]$ and from $[\frac{2}{3}, 1]$ and continue removing the open middle thirds from all remaining intervals. The points remaining form C. Show:

    (a) C is closed,
    (b) $m(C) = 0$,
    (c) the points in C may all be represented in the ternary system (that is, numbers to the base 3) by a decimal with only 0's and 2's,
    (d) C is not denumerable.

14. Let Q be the set of numbers in the interval $[0, 1]$ that do not have a "5" in their decimal expansion. Prove $m(Q) = 0$.

15. If $E_1 \subset E_2 \subset E_3 \subset \cdots$, and $E = \bigcup_{n=1}^{\infty} E_n$, where each E_n $(n = 1, 2, \ldots)$ is measurable, prove that $\lim_{n \to \infty} m(E_n) = m(E)$.

16. If $E_1 \supset E_2 \supset E_3 \supset \cdots$, and $E = \bigcap_{n=1}^{\infty} E_n$, where each E_n $(n = 1, 2, \ldots)$ is measurable, prove that $\lim_{n \to \infty} m(E_n) = m(E)$.

17. Let E be a bounded set of points in $[a, b]$. Prove the following definition is equivalent to Definition 8.2.2:

$$m_{\oplus}(E) = \sup_{F \subset E} m(F),$$

F closed. Discuss extensions of this definition to (a) the case where $[a, b]$ is replaced by a larger interval, and (b) the case where E is unbounded.

IX

LEBESGUE INTEGRABLE FUNCTIONS

9.1 measurable functions

Our purpose in investigating Lebesgue measure and measurable sets was to lay the groundwork for Lebesgue integration. Thus, we want to investigate sets of points in the domain of a function such that the image of these points lies between two numbers, say, A and B. To put it another way, we shall be concerned with the inverse image of the interval (A, B), that is, $\{x \mid A < f(x) < B\}$. Obviously, closed intervals as well as half-open and half-closed intervals must also be considered. Hopefully, these inverse images in the domain of the function will be measurable sets of points, although, as the discussion at the beginning of the last chapter set forth, we should not expect things to be as simple as if we had started with intervals in the domain of definition of the function.

The following definition is thus a reasonable point of departure in constructing a theory of integration based on the more general concept of measure instead of the limited concept of the length of an interval.

9.1.1 *DEFINITION* A function is measurable if any one of the following conditions holds for every real number c:

(a) $S = \{x \mid f(x) > c\}$ is measurable.
(b) $T = \{x \mid f(x) \geq c\}$ is measurable.
(c) $U = \{x \mid f(x) < c\}$ is measurable.
(d) $V = \{x \mid f(x) \leq c\}$ is measurable.

9.1.2 *THEOREM* The conditions of Definition 9.1.1 are equivalent.

Proof (a) implies (b) since

$$\{x \mid f(x) \geq c\} = \bigcap_{n=1}^{\infty} \{x \mid f(x) > c - (1/n)\}.$$

Then use Theorem 8.3.12. The other implications are left as problems. ∎

9.1.3 *COROLLARY* Any of the conditions of Definition 9.1.1 implies that the set $W = \{x \mid f(x) = c\}$ is measurable.

Proof Write $W = T \cap V$ in Definition 9.1.1. ∎

9.1.4 *EXAMPLES*

(a) The function $f(x) = ax^3 + bx^2 + cx + d$ is measurable, for, pick any one of the conditions in Definition 9.1.1, and the set S, T, U, or V is at worst a collection of intervals and isolated points.

(b) The function

$$f(x) = \begin{cases} 1 & \text{if } x \text{ is irrational,} \\ 0 & \text{if } x \text{ is rational,} \end{cases}$$

is measurable; for, let $V = \{x \mid f(x) \leq c\}$. If $0 < c < 1$, V is measurable by Theorems 3.1.5 and 8.2.7. For any other c, the situation is trivial.

The last example shows that the class of measurable functions includes at least some discontinuous functions. The fact that it includes all continuous functions is given in the problems.

It will soon become apparent that measurable functions play about the same role in Lebesgue integration that continuous functions play in Riemann integration. Thus it will be instructive to note the similarities and also the differences between the properties of these two classes of functions. We are not so much concerned with what makes one function continuous and another one only measurable as we are with the properties of each class of functions when considered as a space of elements.

The following theorem shows that the class of measurable functions is closed under addition and scalar multiplication so that, when the other postulates in Definition 1.1.1 are verified it is seen that the class of measurable functions, as well as the class of continuous functions, is a real linear vector space.

9.1.5   *THEOREM*   If f and g are measurable functions and α and β are real numbers, then $\alpha f + \beta g$ is measurable.

> **Proof**   We shall use the set U in Definition 9.1.1. If $\alpha > 0$, we note that $\{x \mid \alpha f(x) < c\} = \{x \mid f(x) < c/\alpha\}$, which shows that αf is measurable when f is. To show that $f + g$ is measurable, let $f(x) + g(x) < c$, and recall from Section 3.1 that there is a rational number r such that
>
> $$f(x) < r < c - g(x).$$
>
> Now, the sets $\{x \mid f(x) < r\}$ and $\{x \mid g(x) < c - r\}$ are both measurable, and therefore their intersection is measurable. Furthermore, $\{x \mid f(x) + g(x) < c\} = \bigcup_r [\{x \mid f(x) < r\} \cap \{x \mid g(x) < c - r\}]$, where the union is taken over the set of all rationals satisfying $f(x) < r < c - g(x)$ for some x. This set is countable so we have the countable union of measurable sets which is measurable by Theorem 8.3.12. To complete the proof, combine the two results above to show that $\alpha f + \beta g$ is measurable.   ∎

9.1.6   *THEOREM*   If f and g are measurable, f^2 and fg are measurable.

> **Proof**   Use the set S in Definition 9.1.1. If $c < 0$,
>
> $$\{x \mid [f(x)]^2 > c\}$$
>
> is the domain of f.

If $c \geq 0$,

$$\{x \mid [f(x)]^2 > c\} = \{x \mid f(x) > \sqrt{c}\} \cup \{x \mid f(x) < -\sqrt{c}\}.$$

In either case, f^2 is measurable. Note that

$$fg = \tfrac{1}{2}[(f+g)^2 - f^2 - g^2],$$

so fg is measurable. ∎

Thus far the class properties of measurable functions parallel those for continuous functions. However, in the remainder of this section significant differences appear.

First, however, recall that lim sup and lim inf of a sequence of real numbers were defined in Section 2.1, and in Problem 14 of that section it was proved that if $\lim \sup_{n \to \infty} a_n = \lim \inf_{n \to \infty} a_n$, then $\lim_{n \to \infty} a_n$ exists and equals the common value of the others. For our present purposes we wish to change the definitions of lim sup and lim inf slightly and ask the student to verify (see Problem 1) that the new definitions are equivalent to those in Definition 2.1.5.

9.1.7 DEFINITION Let $\{f_n\}$ be a sequence of functions. Then for each x,

$$\lim_{n \to \infty} \sup f_n(x) = \inf_n [\sup_{k \geq n} f_k(x)]$$

and

$$\lim_{n \to \infty} \inf f_n(x) = \sup_n [\inf_{k \geq n} f_k(x)].$$

9.1.8 THEOREM Let $\{f_n\}$ be a sequence of measurable functions. Then the functions

(a) $F_1 = \sup(f_1, \ldots, f_n)$,
(b) $F_2 = \inf(f_1, \ldots, f_n)$,
(c) $F_3 = \sup_n\{f_n\}, \ n = 1, 2, \ldots$,
(d) $F_4 = \inf_n\{f_n\}, \ n = 1, 2, \ldots$,
(e) $F_5 = \lim \sup_{n \to \infty} f_n$,
(f) $F_6 = \lim \inf_{n \to \infty} f_n$

are all measurable.

Proof We shall prove (c) and (e) and leave the others as problems. Use set S in Definition 9.1.1. Let $F_3(x) = \sup_n \{f_n(x)\}$. Then,

$$\{x \mid F_3(x) > c\} = \bigcup_{n=1}^{\infty} \{x \mid f_n(x) > c\}$$

so F_3 is measurable. Also, (e) follows from (c) and (d) by the definition of lim sup. ∎

9.1.9 COROLLARY Let $\{f_n\}$ be a sequence of measurable functions. If $\lim_{n \to \infty} f_n(x) = f(x)$ for each x, f is measurable.

Proof This follows from (e) and (f) in Theorem 9.1.8. ∎

Note that we have proved that the *pointwise* limit of a sequence of measurable functions is measurable whereas the pointwise limit of a sequence of continuous functions is not necessarily continuous. It will appear later that this feature is of great importance in convergence properties of Lebesgue integrals.

At this point it might be instructive to ask if there are any functions that are *not* measurable. In fact, contrary to the case of continuous functions, it is difficult to construct a nonmeasurable function. The simplest way is to rely on the nonmeasurable set described in Section 8.3. Let this set be S, and let

$$f(x) = \begin{cases} 1 & \text{if } x \in S, \\ 0 & \text{if } x \notin S. \end{cases}$$

Then f is not measurable, for the set $\{x \mid f(x) > \frac{1}{2}\}$ is not measurable.

PROBLEMS

1. Show that Definitions 2.1.5 and 9.1.7, and also Problem 5, Section 4.2, are consistent formulations of lim sup and lim inf.

2. Complete the proof of Theorem 9.1.2.

3. Prove that a continuous function is measurable.

4. Complete the proof of Theorem 9.1.8.

5. By constructing a counterexample, show that the condition in Corollary 9.1.3 does not imply those in Definition 9.1.1.

6. Give an example of a function f with the property that $|f|$ is measurable but f is not.

7. If g is continuous on $(-\infty, \infty)$ and f is measurable, show that $g \circ f$ is measureable. What can be said about $f \circ g$?

8. If f is not measurable and g is not measurable, what can be said about $f + g, fg,$ and $f \circ g$?

9.2 Lebesgue integral of nonnegative measurable functions

In defining the Riemann integral of f over the interval $[a, b]$, two sums were used whose terms were of the form $m_i \, \Delta x_i$ and $M_i \, \Delta x_i$, where m_i and M_i were constant for x in the ith interval. The m_i and M_i could actually be represented as values of new functions defined as

$$\varphi(x) = \begin{cases} m_1 & \text{if} \quad a \le x < x_1, \\ m_2 & \text{if} \quad x_1 \le x < x_2, \\ \quad \vdots \\ m_n & \text{if} \quad x_{n-1} \le x \le b, \end{cases}$$

and

$$\psi(x) = \begin{cases} M_1 & \text{if} \quad a \le x < x_1, \\ M_2 & \text{if} \quad x_1 \le x < x_2, \\ \quad \vdots \\ M_n & \text{if} \quad x_{n-1} \le x \le b. \end{cases}$$

Another representation is to let

$$\varphi_1(x) = \begin{cases} m_1 & \text{if} \quad a \le x < x_1, \\ 0 & \text{for} \quad \text{all other } x \text{ in } [a, b], \end{cases}$$

$$\varphi_2(x) = \begin{cases} m_2 & \text{if} \quad x_1 \le x < x_2, \\ 0 & \text{for} \quad \text{all other } x \text{ in } [a, b], \end{cases}$$

$$\vdots$$

$$\varphi_n(x) = \begin{cases} m_n & \text{if} \quad x_{n-1} \leq x \leq b, \\ 0 & \text{for} \quad \text{all other } x \text{ in } [a, b]. \end{cases}$$

Then, $\varphi(x) = \sum_{k=1}^{n} \varphi_k(x)$.

Similar definitions will give $\psi(x)$.

The functions φ and ψ are called *step functions* because their values are constant for x in intervals.

In defining the Lebesgue integral the following procedural changes are made from the Riemann case:

(1) The intervals $[x_{i-1}, x_i]$ will be replaced by measurable sets E_i, and the lengths Δx_i will be replaced by $m(E_i)$, $i = 1, 2, \ldots, n$.

(2) The functions φ and ψ must be constant over the measurable sets E_i instead of over intervals $[x_{i-1}, x_i]$. They are then called *simple functions* (definition follows).

(3) We shall define the Lebesgue integral of a nonnegative simple function first and then use this integral to define the integral of a nonnegative bounded measurable function over a set of finite measure. Contrary to the Riemann case, this integral will always exist. When the conditions of being nonnegative and bounded over a set of finite measure are relaxed, the integral will still exist if we are careful to rule out the cases where $\int f = \infty$. Thus we will be able to handle the integral of every measurable function.

9.2.1 DEFINITION

(a) For each set E, the *characteristic function* χ_E is defined as

$$\chi_E(x) = \begin{cases} 1 & \text{if} \quad x \in E, \\ 0 & \text{if} \quad x \notin E. \end{cases}$$

(b) A *simple function* is a function of the form $\varphi(x) = \sum_{i=1}^{n} c_i \chi_{E_i}(x)$, where the sets E_i are measurable.

As we shall use the term "simple function," the set of values $\{c_i\}$ is finite, although some authors allow a countably infinite number of different values. The latter case will be referred to here as an *elementary function*.

Note that the representation in (b) is not unique. We may make it so by choosing the representation that is "best" in the following sense: let the finite set $\{c_i, \ldots, c_n\}$ be given first, $c_i \neq 0$, $c_i \neq c_j$ if $i \neq j$, and let $E_i = \{x \mid \varphi(x) = c_i\}$. Then the E_i are disjoint and measurable, and $\varphi = \sum_{i=1}^n c_i \chi_{E_i}$ is unique. We shall refer to this as the *standard form* of a simple function.

9.2.2   DEFINITION   Let $\varphi = \sum_{i=1}^n c_i \chi_{E_i}$ be a simple function in standard form defined over a set E, where $\bigcup_{i=1}^n E_i \subset E$, $m(E_i) < \infty$ $(i = 1, \ldots, n)$ and $\varphi(x) = 0$ for $x \in E - \bigcup_{i=1}^n E_i$. Then the *Lebesgue integral* of φ over E is

$$\int_E \varphi(x) \, dx = \sum_{i=1}^n c_i m(E_i).$$

It may be necessary at times to distinguish between the Riemann integral and the Lebesgue integral. In these cases we may use R $\int$ and L $\int$. From this point on, when the integral sign is used alone it will stand for the Lebesgue integral. If the set E is an interval, the familiar notation $\int_a^b$ will be used; and, when no confusion can result, references to the set E and the dummy variable x may be omitted.

It should be obvious at this point that the value of the integral does not depend on the form (standard or otherwise) of φ.

In case the simple function φ is actually a step function, it is easily seen that Definition 9.2.2 reduces to just the sum of the areas of certain rectangles. For example, if $n = 4$, we might have the rectangles shown in Fig. 10.

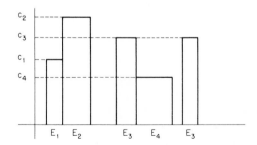

Figure 10

9.2.3 *THEOREM* Let φ and ψ be simple functions with the notation of Definition 9.2.2. Then, if α and β are real numbers

(1) $\int (\alpha\varphi + \beta\psi) = \alpha \int \varphi + \beta \int \psi.$

(2) If $\varphi \le \psi$ except possibly on a set of measure zero,

$$\int \varphi \le \int \psi$$

Proof We shall start with the functions φ and ψ in standard form defined on sets in the sequences $\{E_i'\}$ and $\{E_j''\}$, respectively. Let $E_k = E_i' \cap E_j''$, where all possible intersections of sets in $\{E_i'\}$ and $\{E_j''\}$ are taken to form the sets in $\{E_k\}$. As noted, the values of the integrals of φ and ψ are not changed by replacing some E_i' and E_j'' by (probably several) sets in the sequence $\{E_k\}$. Thus,

$$\varphi = \sum_{k=1}^{m} c_k \chi_{Ek}, \qquad \psi = \sum_{k=1}^{m} b_k \chi_{Ek}$$

and

$$\alpha\varphi + \beta\psi = \sum_{k=1}^{m} (\alpha c_k + \beta b_k) \chi_{Ek}$$

$$= \alpha \sum_{k=1}^{m} c_k \chi_{Ek} + \beta \sum_{k=1}^{m} b_k \chi_{Ek}.$$

Then part (1) follows immediately. Part (2) follows by noting that $c_k \le b_k$ for $k = 1, \ldots, m$. ∎

We are now in a position to prove a fundamental theorem that will assure the existence of the Lebesgue integral of a nonnegative bounded measurable function over a set of finite measure.

9.2.4 *THEOREM* Let f be a nonnegative bounded measurable function on a set E with $m(E) < \infty$. Let $\varphi \le f$ and $\psi \ge f$ be simple functions. Then,

$$\sup_{\varphi} \int_E \varphi = \inf_{\psi} \int_E \psi,$$

where the sup is taken over all simple functions $\varphi \leq f$ and the inf is taken over all simple functions $\psi \geq f$.

Proof   Let $0 \leq f(x) < M$ for $x \in E$, and for each n and each $k = 1, 2, \ldots, n$, let

$$E_k = \{x \mid \frac{(k-1)}{n} M \leq f(x) \leq \frac{k}{n} M\}.$$

Then, $\bigcup_{k=1}^{n} E_k = E$, $E_i \cap E_j = \phi$ if $i \neq j$, and E_k is measurable for $k = 1, \ldots, n$. Thus $\sum_{k=1}^{n} m(E_k) = m(E)$. Now, form the simple functions φ and ψ, where

$$\varphi_n(x) = \frac{M}{n} \sum_{k=1}^{n} (k-1)\chi_{E_k}(x)$$

and

$$\psi_n(x) = \frac{M}{n} \sum_{k=1}^{n} k\chi_{E_k}(x).$$

Obviously, $\varphi_n \leq f \leq \psi_n$. Now,

$$\int_E \varphi_n(x)\, dx = \frac{M}{n} \sum_{k=1}^{n} (k-1)m(E_k)$$

$$= \frac{M}{n} \sum_{k=1}^{n} km(E_k) - \frac{M}{n} \sum_{k=1}^{n} m(E_k)$$

$$= \frac{M}{n} \sum_{k=1}^{n} km(E_k) - \frac{M}{n} m(E).$$

and

$$\int_E \psi_n(x)\, dx = \frac{M}{n} \sum_{k=1}^{n} km(E_k)$$

So,

$$\int_E \psi_n(x) \ dx - \int_E \varphi_n(x) \ dx = \frac{M}{n} \ m(E).$$

But, with the sup and inf described in the theorem,

$$\inf \int_E \psi(x) \ dx \le \int_E \psi_n(x) \ dx$$

and

$$\sup \int_E \varphi(x) \ dx \ge \int_E \varphi_n(x) \ dx.$$

Thus,

$$\inf \int \psi - \sup \int \varphi \le \frac{M}{n} \ m(E).$$

But, since n is arbitrary,

$$\sup_{\varphi \le f} \int \varphi = \inf_{\psi \ge f} \int \psi. \quad \blacksquare$$

9.2.5 *DEFINITION* If f is a nonnegative bounded measurable function defined on a measurable set E with $m(E) < \infty$, the Lebesgue integral of f over E is the common value of $\sup_{\varphi \le f} \int \varphi$ and $\inf_{\psi \ge f} \int \psi$, where φ and ψ are simple functions.

$$\int_E f(x) \ dx = \sup_{\varphi \le f} \int_E \varphi(x) \ dx = \inf_{\psi \ge f} \int_E \psi(x) \ dx.$$

Thus the integral of a measurable function f over a measurable set E always exists as a finite number under the following conditions:

(1) f is bounded,
(2) $m(E)$ is finite,
(3) $f \ge 0$.

We shall relax these conditions shortly, but in defining the Lebesgue integral in more general cases we need to make use of certain properties of the integral already defined. We shall state these properties as a theorem here in more detail than is necessary merely to justify the definition that follows; however, it is later easy to note that these same properties hold in the general case of integrable functions.

The following theorem also shows that in many cases we are not concerned with what happens to a function on sets of measure zero. If a certain condition holds everywhere except on a set of measure zero, we say it holds *almost everywhere*, or a.e. Thus, if $f = g$ a.e., we mean that $f(x) = g(x)$ for all $x \in E$, where $m(CE) = 0$.

9.2.6 THEOREM Let f and g be nonnegative bounded measurable functions defined on measurable sets $E, E_1 E_2$, of finite measure. Then

(a) if $\alpha, \beta \geq 0$, $\displaystyle\int_E (\alpha f + \beta g) = \alpha \int_E f + \beta \int_E g$,

(b) if $m(E) = 0$, $\displaystyle\int_E f = 0$,

(c) if $f = g$ a.e., $\displaystyle\int_E f = \int_E g$,

(d) if $f \leq g$ a.e., $\displaystyle\int_E f \leq \int_E g$,

(e) if $E_1 \subset E_2$, $\displaystyle\int_{E_1} f \leq \int_{E_2} f$,

(f) if $A \leq f(x) \leq B$ a.e., $Am(E) \leq \displaystyle\int_E f \leq Bm(E)$,

(g) if $E_1 \cap E_2 = \phi$, $\displaystyle\int_{E_1 \cup E_2} f = \int_{E_1} f + \int_{E_2} f$.

Proof Let φ and ψ be simple functions. Then $\alpha\varphi$ is a simple function and, since $\alpha \geq 0$,

$$\int \alpha f = \sup_{\alpha\varphi \le \alpha f} \int \alpha\varphi = \alpha \sup_{\varphi \le f} \int \varphi = \alpha \int f$$

Now, let $\varphi_1 \le f \le \psi_1$ and $\varphi_2 \le f \le \psi_2$. Then,

$$\int \varphi_1 + \int \varphi_2 = \int (\varphi_1 + \varphi_2) \le \int (f + g) \le \int (\psi_1 + \psi_2)$$

$$= \int \psi_1 + \int \psi_2.$$

But,

$$\int f = \sup_{\varphi_1 \le f} \int \varphi_1 = \inf_{f \le \psi_1} \int \psi_1$$

and

$$\int g = \sup_{\varphi_2 \le g} \int \varphi_2 = \inf_{g \le \psi_2} \int \psi_2,$$

so,

$$\int f + \int g \le \int (f + g) \le \int f + \int g,$$

and part (a) follows by combining the above results. Proofs of the other parts are left as problems with the suggestions for (d) and (f) that if $A \subset B$, $\int_A f = \int_B f \cdot \chi_A$, and if $A \cap B = \phi$, $\chi_{A \cup B} = \chi_A + \chi_B$. ∎

Let us now return to the problem of defining $\int_E f$ in case f is unbounded and/or $m(E) = \infty$. The method is somewhat similar to that used in defining improper Riemann integrals but with certain distinctions. First, as long as the function is nonnegative, there is no need to treat the "infinite pieces" separately; and second, the difficulty that arises by having the function unbounded is treated by cutting off the unbounded portion by horizontal lines instead of vertical lines as shown in Fig. 11.

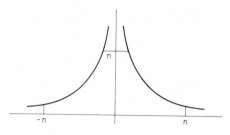

Figure 11

9.2.7 DEFINITION Let f be nonnegative and measurable on E, where E has the property that if I is any bounded interval, $E \cap I$ is measurable and $m(E \cap I) < \infty$. Let F_n be defined by

$$F_n(x) = \begin{cases} f(x) & \text{if} \quad |x| \leq n, \quad f(x) \leq n, \\ n & \text{if} \quad |x| \leq n, \quad f(x) > n, \\ 0 & \text{if} \quad |x| > n. \end{cases}$$

Then

$$\int_E f = \lim_{n \to \infty} \int_E F_n.$$

We should observe that in a sense the above definition approximates the function f by hybrid functions F_n that are part simple (the last two lines in the definition) and part just measurable. Two other approaches are possible, however, since we could have approximated f entirely by simple functions or entirely by bounded measurable functions. These approaches are suggested in the problems.

Since $\{ \int_E F_n \}$ in the above definition is a nondecreasing sequence of positive numbers, $\int_E f$ always exists if we allow the value $+\infty$. An unfortunate semantic difficulty exists in the literature, however, because we may write $\int_E f = \infty$, but we do not say f is *integrable* unless $\int_E f < \infty$.[†]

[†] Some authors use the word *summable* in this situation, and the student is advised to check the definitions carefully in consulting other references.

9.2.8   *THEOREM*   Parts (a)–(g) of theorem 9.2.6 hold for non-negative measurable functions.

Proof   As in the former proof we shall prove only (a) in detail.

$$\int \alpha f = \lim_{n \to \infty} \int \alpha F_n = \alpha \lim_{n \to \infty} \int F_n = \alpha \int f.$$

Also, let f and g be approximated by $\{F_n\}$ and $\{G_n\}$, respectively, as in Definition 9.2.7. Then, if $h = f + g$, let h be approximated in the same way, so

$$H_n(x) \leq F_n(x) + G_n(x) \leq H_{2n}(x).$$

Therefore,

$$\int H_n \leq \int (F_n + G_n) \leq \int H_{2n}$$

or

$$\int H_n \leq \int F_n + \int G_n \leq \int H_{2n}.$$

So, taking the limit as $n \to \infty$ we have

$$\int h \leq \int f + \int g \leq \int h$$

and the result follows.   ∎

9.2.9   *COROLLARY*   If f and g are nonnegative integrable functions on E and $g \leq f$, then

$$\int_E (f - g) = \int_E f - \int_E g.$$

Proof   Write $f = (f - g) + g$.   ∎

PROBLEMS

1.   Complete the proof of Theorem 9.2.6.

2.   Complete the proof of Theorem 9.2.8.

3.   Let f be a nonnegative measurable function (possibly unbounded) defined on a measurable set E [possibly $m(E) = \infty$]. Let g represent a bounded nonnegative measurable function such that $m(\{x \mid g(x) \neq 0\}) < \infty$. Show that the following definition is equivalent to Definition 9.2.7.

$$\int_E f = \sup_{g \leq f} \int_E g.$$

4.   Let f be a nonnegative measurable function (possibly unbounded) defined on a measurable set E [possibly $m(E) = \infty$]. Let $E_1 \subset E_2 \subset E_3 \subset \cdots \subset E_n \subset \cdots$, with $m(E_n) < \infty$ for each n, and $\bigcup_{n=1}^{\infty} E_n = E$. Let

$$\varphi_n(x) = \begin{cases} \dfrac{k}{2^n} & \text{if } \ x \in E_n \ \text{ and } \ \dfrac{k}{2^n} \leq f(x) < \dfrac{k+1}{2^n}, \\[2mm] & \qquad k = 0, 1, 2, \ldots, (n2^n - 1), \\[2mm] n & \text{if } \ x \in E_n, \ f(x) \geq n, \\[2mm] 0 & \text{if } \ x \notin E_n. \end{cases}$$

Show that $\{\varphi_n\}$ is a nondecreasing sequence of simple functions converging to f everywhere on E.

5.   Let φ_n be defined as in Problem 4 for $n = 1, 2, \ldots$. Show that $\int_E f = \lim_{n \to \infty} \int_E \varphi_n$ is equivalent to Definition 9.2.7.

6.   Let

$$R \underline{\int_a^b} f(x) \, dx \qquad \text{and} \qquad R \overline{\int_a^b} f(x) \, dx$$

be the lower and upper Riemann integrals in Remark 6.1.1 (f), and let φ and ψ be simple functions as in Theorem 9.2.4. Show that

$$R \underline{\int_a^b} f(x) \, dx \leq \sup \int_a^b \varphi(x) \, dx \leq \inf \int_a^b \psi(x) \, dx \leq R \overline{\int_a^b} f(x) \, dx.$$

Then show that if f is Riemann integrable on $[a, b]$, f is Lebesgue integrable on $[a, b]$ and the two integrals are equal.

7. Prove that f is Riemann integrable on $[a, b]$ if and only if the
    set of points in $[a, b]$, where f is discontinuous, has Lebesgue
    measure zero.

9.3 Lebesgue   integrable   functions—convergence
properties

In this section we shall prove three convergence theorems for
Lebesgue integrals. First, however, it is necessary to define the integral without the restriction that $f \geq 0$.
   We define f^+ and f^- as in Remark 6.1.1 so that $f^+(x) = \max(f(x), 0)$, $f^-(x) = \max(-f(x), 0)$. $f = f^+ - f^-$, and $|f| = f^+ + f^-$.

9.3.1 *DEFINITION* Let f be measurable on E. If $\int_E f^+ < \infty$ and
$\int_E f^- < \infty$, f is integrable on E and

$$\int_E f = \int_E f^+ - \int_E f^-.$$

In case $\int_E f^+ = \infty$ and $\int_E f^- < \infty$, we shall write $\int_E f = \infty$; but, as before, f is not *integrable* unless $\int_E f < \infty$. (A
similar statement holds in case $\int_E f^- = \infty$ and $\int_E f^+ < \infty$.
We obviously do not allow $\infty - \infty$.)

9.3.2 *THEOREM* Parts (a)–(g) of Theorem 9.2.6 hold for
integrable functions, and, in addition, $(h) \; |\int_E f| \leq \int_E |f|$.

Proof For (a), note that Corollary 9.2.9 can be applied to
$f^+ - f^-$ and $f^- - f^+$ on sets where they are nonnegative. The
rest of the proof is left as problems. ∎

9.3.3 *EXAMPLES*

   (a) We have $f(x) = 1/2\sqrt{x}$ integrable on $(0, 10)$ since

$$F_n(x) = \begin{cases} \dfrac{1}{2\sqrt{x}} & \text{for} \quad \dfrac{1}{2\sqrt{x}} \le n, \quad |x| \le n, \\[3mm] n & \text{for} \quad \dfrac{1}{2\sqrt{x}} > n, \quad |x| \le n, \\[3mm] 0 & \text{for} \quad |x| > n. \end{cases}$$

If values of F_n are restated for domains in terms of x, we have

$$F_n(x) = \begin{cases} \dfrac{1}{2\sqrt{x}} & \text{for} \quad \dfrac{1}{4n^2} \le x \le \min(10, n), \\[3mm] n & \text{for} \quad 0 < x < \dfrac{1}{4n^2}, \\[3mm] 0 & \text{for} \quad x > \min(10, n). \end{cases}$$

Then,

$$\lim_{n \to \infty} \int_0^n F_n(x)\, dx = \lim_{n \to \infty} \left[\int_0^{1/4n^2} n\, dx + \int_{1/4n_2}^{10} \frac{1}{2\sqrt{x}}\, dx \right].$$

The first integral is the integral of a simple function and its value is $(1/4n^2)n = 1/4n$. By Problem 6, Section 9.2, the second integral may be evaluated as a Riemann integral and is found to be $\sqrt{10} - (1/2n)$. Thus,

$$\int_0^{10} f(x)\, dx = \sqrt{10}.$$

(b) We have $f(x) = 1/x^2$ integrable on $E = (\tfrac{1}{2}, \infty)$ since

$$F_n(x) = \begin{cases} \dfrac{1}{x^2} & \text{for} \quad \dfrac{1}{x^2} \le n, \quad |x| \le n, \\[3mm] n & \text{for} \quad \dfrac{1}{x^2} > n, \quad |x| \le n, \\[3mm] 0 & \text{for} \quad |x| > n. \end{cases}$$

Again, restating this for domain in terms of x,

$$F_n(x) = \begin{cases} \dfrac{1}{x^2} & \text{for} \quad \max\!\left(\dfrac{1}{\sqrt{n}},\dfrac{1}{2}\right) \le x \le n, \\[3mm] n & \text{for} \quad \dfrac{1}{2} < x < \dfrac{1}{\sqrt{n}}, \\[3mm] 0 & \text{for} \quad x > n. \end{cases}$$

Now, since the middle line is meaningless for $n > 4$,

$$\int_{1/2}^{\infty} f(x)\,dx = \lim_{n\to\infty}\left[\int_{1/2}^{n}\frac{1}{x^2}\,dx + \int_{n}^{\infty} 0\,dx\right] = 2.$$

(c) Example 6.3.2, which was given for Riemann integrals, holds just as well for Lebesgue integrals. That is,

$$f(x) = \frac{d}{dx}\!\left(x^2 \sin\frac{1}{x^2}\right)$$

is not Lebesgue integrable on $[0, 1]$ for the same reasons cited in the example.

In the previous examples we found the value of a Lebesgue integral by noting that in these cases it was also a Riemann integral and then evaluating it by elementary means that go back to the fundamental theorem of integral calculus.

A similar procedure can be justified for the Lebesgue integral although the development is somewhat tedious due to the presence of sets of measure zero. The main results are that if f is Lebesgue integrable on $[a, b]$ and

$$F(x) = \int_{a}^{x} f(t)\,dt \qquad (a < x < b),$$

then $F'(x) = f(x)$ a.e. on $[a, b]$.

If we start with a function F that is differentiable at *every* point of $[a, b]$, then

$$F(x) - F(a) = \int_{a}^{x} F'(t)\,dt \qquad (a < x < b).$$

9.3.4 *THEOREM* If $\{f_n\}$ is a nondecreasing sequence of non-negative integrable functions and $\lim_{n \to \infty} f_n(x) = f(x)$ for all $x \in E$, then

$$\lim_{n \to \infty} \int_E f_n = \int_E f.$$

(*Note:* $\int_E f_n < \infty$ for $n = 1, 2, \ldots$, but $\int_E f$ may be infinite.)

Proof For every n, let $\{\varphi_{n,k}\}$, $k = 1, 2, \ldots$, be a nonde-creasing sequence of nonnegative simple functions converging to f_n. (This can be accomplished by first taking the approxi-mating functions in Definition 9.2.7 and replacing the bounded portion by simple functions which converge to those bounded portions as in the proof of Theorem 9.2.4.) Let $\Phi_n = \max\{\varphi_{i,j}\}$, $i = 1, \ldots, n$, $j = 1, \ldots, n$. Then $\Phi_n(x) \le f_n(x)$ for every $x \in E$, so

$$\int_E \Phi_n \le \int_E f_n.$$

Now, $\{ \int_E \Phi_n \}$ and $\{ \int_E f_n \}$ are nonnegative and nonde-creasing, so $\lim_{n \to \infty} \int_E \Phi_n$ and $\lim_{n \to \infty} \int_E f_n$ both exist (pos-sibly infinite) and

$$\lim_{n \to \infty} \int_E \Phi_n \le \lim_{n \to \infty} \int_E f_n.$$

Now, let $p \ge \max(n, k)$ for every n and k. Then $\Phi_p(x) \ge \varphi_{n,k}(x)$ for $x \in E$, and thus

$$\lim_{p \to \infty} \int_E \Phi_p \ge \lim_{k \to \infty} \int_E \varphi_{n,k} = \int_E f_n.$$

Hence,

$$\lim_{n \to \infty} \int_E \Phi_n = \lim_{n \to \infty} \int_E f_n.$$

But, $\{\Phi_p\}$ is a sequence of simple functions such that $\Phi_p \ge f_n$, $n = 1, 2, \ldots$, $p \ge n$. That is, $f_n(x) \le \Phi_p(x) \le f(x)$ for all $x \in E$, $p \geq n$. Thus, $\lim_{n \to \infty} \int_E \Phi_n = \int_E f$. Combining this with the above,

$$\lim_{n \to \infty} \int_E f_n = \int_E f. \quad \blacksquare$$

The condition that f_n be nonnegative can easily be removed as shown:

9.3.5 *COROLLARY* If $\{f_n\}$ is a nondecreasing sequence of integrable functions and $\lim_{n \to \infty} f_n(x) = f(x)$ for all $x \in E$, then,

$$\lim_{n \to \infty} \int_E f_n = \int_E f.$$

Proof We have $\{f_n - f_1\}$ as a nondecreasing sequence of nonnegative integrable functions which converges to $f - f_1$. Then apply Theorem 9.3.4. $\blacksquare$

9.3.6 *THEOREM* (*Fatou's Lemma*) Let $\{f_n\}$ be a sequence of nonnegative integrable functions on E. Then

$$\int_E \liminf_{n \to \infty} f_n \leq \liminf_{n \to \infty} \int_E f_n.$$

Proof Note that $\liminf_{n \to \infty} f_n$ and $\liminf_{n \to \infty} \int_E f_n$ always exist. (If $\liminf_{n \to \infty} f_n = +\infty$, the theorem is trivially true.) Let $g_n = \inf\{f_n, f_{n+1}, \ldots\}$, $n = 1, 2, \ldots$. Then $\lim_{n \to \infty} g_n = \liminf_{n \to \infty} f_n$, and $\{g_n\}$ is nondecreasing. Thus,

$$\int_E (\liminf_{n \to \infty} f_n) = \int_E \lim_{n \to \infty} g_n = \lim_{n \to \infty} \int_E g_n$$

by the previous theorem. But, $g_n(x) \leq f_n(x)$ for every $x \in E$, so $\int_E g_n \leq \int_E f_n$ and thus

$$\lim_{n \to \infty} \int_E g_n \leq \liminf_{n \to \infty} \int_E f_n.$$

Combining this with the preceding equality gives the result. $\blacksquare$

If, in Theorem 9.3.6, we know that $\lim_{n \to \infty} f_n(x) = f(x)$ for all $x \in E$ or even just a.e. in E, the conclusion of the theorem can be stated

$$\int_E f \leq \liminf_{n \to \infty} \int_E f_n.$$

9.3.7 COROLLARY   If $\{f_n\}$ is a sequence of nonpositive integrable functions, then

$$\int_E (\limsup_{n \to \infty} f_n) \geq \limsup_{n \to \infty} \int_E f_n.$$

Proof   The proof is left as Problem 1. ∎

9.3.8 THEOREM   (*Lebesgue convergence theorem*) Let $\{f_n\}$ be a sequence of integrable functions such that $\lim_{n \to \infty} f_n(x) = f(x)$ a.e. in E and $|f_n(x)| \leq g(x)$, where g is integrable on E. Then

$$\lim_{n \to \infty} \int_E f_n = \int_E f.$$

Proof   The function $g - f_n$ is nonnegative so, by Theorem 9.3.6,

$$\int_E g - f \leq \liminf_{n \to \infty} \int_E (g - f_n).$$

Since $|f_n| \leq g$, $|f| \leq g$, and so f is integrable. Thus,

$$\int_E g - \int_E f \leq \int_E g - \limsup_{n \to \infty} \int_E f_n,$$

so

$$\int_E f \geq \limsup_{n \to \infty} \int_E f_n.$$

Similarly, $g + f_n$ is nonnegative, and, by the same procedure as above,

$$\int_E f \le \liminf_{n \to \infty} \int_E f_n.$$

Thus,

$$\int_E f \le \liminf_{n \to \infty} \int_E f_n \le \limsup_{n \to \infty} \int_E f_n \le \int_E f,$$

so

$$\lim_{n \to \infty} \int_E f_n = \int_E f. \quad \blacksquare$$

The preceding theorem is sometimes called the *dominated convergence theorem*; if $g(x) = M$ (a constant) and $m(E) < \infty$, it is called the *bounded convergence theorem*. It is one of the most powerful tools in Lebesgue integration. Special care must be taken in its application when $g(x)$ is not a constant, when $m(E) = \infty$, and when the functions take on both positive and negative values. It should be emphasized in applying Theorem 9.3.8 that all functions in the sequence $\{f_n\}$ must be integrable and g must be integrable.

9.3.9 EXAMPLES

(a) Let

$$f_n(x) = \begin{cases} 0 & \text{if } 0 \le x \le n, \\ 1 & \text{if } x > n. \end{cases}$$

Then, $\lim_{n \to \infty} f_n(x) = 0$ (in fact, the convergence is uniform) but

$$\lim_{n \to \infty} \int_0^\infty f_n \ne \int_0^\infty \lim_{n \to \infty} f_n.$$

It should be observed, however, that f_n is not integrable on $[0, \infty]$ for any n, and of course no integrable g with $|f| \le g$ can be found, so Theorem 9.3.8 does not apply.

(b) Let

$$f_n(x) = \lim_{k \to \infty} (\cos n!\pi x)^{2k} \quad \text{and} \quad f(x) = \lim_{n \to \infty} f_n(x).$$

Then, f_n is integrable, $n = 1, 2, \ldots$, and $g(x)$ in Theorem 9.3.8 may be taken as identically 1. Thus f is integrable and

$$\lim_{n \to \infty} \int_0^1 f_n = \int_0^1 \lim_{n \to \infty} f_n = 1.$$

(See Problem 2.)

PROBLEMS

1. Prove Corollary 9.3.7.
2. Show that f in Example 9.3.9 (b) is the same as f in Example 9.1.4 (b).

For Problems 3–8, graphs in the Appendix may help describe the analytical conclusions requested.

3. Show that Theorem 9.3.8 applies as the *bounded convergence theorem* to $\{f_n\}$, where $f_n(x) = nx/(1 + n^2x^2)$, $E = [0, 1]$.
4. Show that Theorem 9.3.8 applies as the *dominated convergence theorem* to $\{f_n\}$, where $f_n(x) = n^{3/2}x/(1 + n^2x^2)$, $E = [0, 1]$.
5. Show that Theorem 9.3.8 does not apply to $\{f_n\}$, where $f_n(x) = n^2x/(1 + n^2x^2)$, $E = [0, 1]$.
6. Do Problem 3 for $f_n(x) = nx(1 - x)^n$.
7. Do Problem 4 for $f_n(x) = n^{3/2}x(1 - x)^n$.
8. Do Problem 5 for $f_n(x) = n^2x(1 - x)^n$.
9. If $f_n(x) = nxe^{-nx^2}$, show that the "$<$" sign holds in Fatou's lemma.
10. Show that

$$\left(1 - \frac{t}{n}\right)^n < e^{-t} \qquad \text{and} \qquad \lim_{n \to \infty} \left(1 - \frac{t}{n}\right)^n = e^{-t}.$$

Prove that

$$\lim_{n \to \infty} \int_0^n \left(1 - \frac{t}{n}\right)^n t^{x-1}\, dt = \int_0^\infty e^{-t}t^{x-1}\, dt.$$

(This is the gamma function of Example 6.3.3. The proof in this problem is a major step in writing the gamma function as an infinite product (see Problem 36, Section 2.4). For a full discussion see the work of Whittaker and Watson.[†]

11.   If $f_n(x) = ae^{-anx} - be^{-bnx}$ $(0 < a < b)$, prove that

$$\sum_{n=1}^{\infty} \int_0^{\infty} f_n(x)\ dx \ne \int_0^{\infty} \left[\sum_{n=1}^{\infty} f_n(x)\right] dx.$$

Verify directly that $\sum_{n=1}^{\infty} \int_0^{\infty} |f_n(x)|\ dx$ diverges.

12.   Show that

$$\int_0^1 \frac{x^p}{1-x} \ln\left(\frac{1}{x}\right) dx = \frac{1}{(p+1)^2} + \frac{1}{(p+2)^2} + \cdots \quad (p > -1).$$

13.   Show that

$$\int_0^{\infty} \frac{\sin ax}{e^x - 1}\ dx = \pi\left(\frac{1}{e^{2\pi a} - 1} - \frac{1}{2\pi a} + \frac{1}{2}\right).$$

9.4   the space $\mathcal{L}_p$

The preceding section showed that under certain reasonable conditions that control the magnitude of $|f(x)|\ m(E)$, $x \in E$, the measurable function f is integrable and that the pointwise limit of a sequence of integrable functions is integrable. Since measurability is not a very restrictive condition (we have noted the difficulty of even demonstrating the existence of a nonmeasurable function in Section 9.1), it should be hoped that most functions encountered in applications will be integrable and that convergence problems will be a great deal simpler than when using Riemann integration. Such is indeed the case.

In order to discuss convergence problems we shall introduce a norm in the linear space of integrable functions and prove that the space is complete under this norm. Note that if f is measurable,

[†] E. T. Whittaker and G. N. Watson, "A Course of Modern Analysis," American ed., Chapter XII, pp. 235–264. Cambridge Univ. Press, London and New York, 1943.

f^p $(1 \leq p < \infty)$ is measurable. However, if f is integrable on E (that is, if $\int_E |f| < \infty$), f^p may or may not be integrable on E; and, conversely, if f^p is integrable on E for some $p > 1$, there is no assurance that f is integrable on E. We therefore introduce normed spaces L_p for every p, $1 \leq p < \infty$, whose elements are functions $f, g, \ldots$ such that the pth powers are integrable.

In constructing a norm in L_p the first property of $\|\cdot\|$ in Definition 1.2.1 causes some slight difficulty. If f is integrable, $|f|$ is integrable, so $\|f\|_p = \int_E |f|^p$ is a natural definition; this is nonnegative, but $\int_E |f|^p$ may be zero without having $f = 0$. If $\int_E |f| = 0$, we know that $f = 0$ a.e., so we can avoid the difficulty by agreeing not to distinguish between two functions that are equal except on a set of measure zero. To this end define f and g to be *equivalent* if $f = g$ a.e., and we shall then deal with classes of equivalent functions. Instead of inventing new symbols for these classes we shall let any function in the class represent the class of equivalent functions, but we shall change the symbol L_p to $\mathcal{L}_p$. Thus the elements of $\mathcal{L}_p$ are classes of equivalent functions.

9.4.1 *NOTATION* The linear space $\mathcal{L}_p$ consists of classes of equivalent functions, where f is equivalent to g if and only if $f = g$ a.e. Any function may represent its equivalence class. If f^p is Lebesgue integrable on E, that is, f is Lebesgue measurable on E and $\int_E |f|^p < \infty$, $f \in \mathcal{L}_p(E)$. Then $\|f\|_p = \int_E |f|^p$.

In the above notation, reference to the set E may be omitted.

9.4.2 *THEOREM* Hölder's inequality and Minkowski's inequality (Theorems 1.3.5 and 1.3.6) hold for elements of $\mathcal{L}_p$.

Proof The proofs given for Theorems 1.3.5 and 1.3.6 carry over directly to the case of Lebesgue integrable functions when a_k is replaced by $f(t)$, b_k is replaced by $g(t)$, $\sum |a_k|^p$ becomes $\int_E |f|^p$, and $\sum |b_k|^p$ becomes $\int_E |g|^p$. ∎

9.4.3 *THEOREM* The symbol $\|f\|_p$ defined in Notation 9.4.1 is a norm according to Definition 1.2.1.

Proof We have already discussed part (1) of Definition 1.2.1; part (2) depends on Theorem 9.4.2; part (3) is a property of Lebesgue integrals (see problem 2). ∎

We now come to one of the most important theorems in modern analysis, a result that should have been anticipated by our motivation to find something "better" than Riemann integration.

9.4.4 *THEOREM* (*Riesz–Fischer theorem*) The linear space $\mathcal{L}_p$ is complete under the norm

$$\| f \|_p = \int_E | f |^p.$$

Proof Let $\{f_n\}$ be a Cauchy sequence in $\mathcal{L}_p$; that is, $f_n \in \mathcal{L}_p$, $n = 1, 2, \ldots$, and $\| f_n - f_m \| \to 0$ as $m, n \to \infty$. We wish to show there exists a function $f \in \mathcal{L}_p$ such that $\| f_n - f \| \to 0$ as $n \to \infty$. Two points should be made before we produce the function f.

(1) We shall select a subsequence of $\{f_n\}$ that converges to a function f and show $f \in \mathcal{L}_p$. This is sufficient to show $\| f_n - f \|_p \to 0$ (see Problem 11, Section 3.1).

(2) Since we are dealing with function in $\mathcal{L}_p$, we are really discussing classes of equivalent functions, so the convergence of the subsequence described in (1) need only be a.e. to f before we show norm convergence.

Now, since $\{f_n\}$ is a Cauchy sequence, there is an N_1 such that $\| f_n - f_m \|_p < \frac{1}{2}$ when $m, n > N_1$; also there is an N_2 such that $\| f_n - f_m \|_p < 1/2^2$ if $m, n > N_2$; there is an N_3 such that $\| f_n - f_m \|_p < 1/2^3$ if $m, n > N_3$; and so on. Now, let $g_1 = f_{N_1+1}, g_2 = f_{N_2+1}, g_3 = f_{N_3+1}, \ldots$. Then $\{g_n\}$ is a subsequence of $\{f_n\}$, and we have avoided the use of double subscripts suggested for Problem 11, Section 3.1. Let $h_1 = | g_1 - g_2 |, h_2 = | g_2 - g_3 |, \ldots$. Then $\| h_n \|_p = \| g_n - g_{n+1} \|_p < 1/2^n$. So,

$$\| h_1 + h_2 + \cdots + h_n \|_p \le \frac{1}{2} + \frac{1}{2^2} + \cdots + \frac{1}{2^n} < 1$$

by the triangle inequality. Now, let

$$s_n(x) \; = \; \sum_{k=1}^{n} h_k(x).$$

Then, $\{s_n\}$ is a nondecreasing sequence of nonnegative integrable functions with $\| s_n \|_p < 1$.

Let $F = \lim_{n \to \infty} s_n$, so by Theorem 9.3.4 $\lim_{n \to \infty} \int s_n = \int F \geq 0$. But, since $\| s_n \|_p < 1$, $\int F \leq 1$ and so $\int F^p \leq 1$. Therefore $\sum_{k=1}^{\infty} | g_k - g_{k+1} |$ is finite a.e., and so is $\sum_{k=1}^{\infty} [g_k(x) - g_{k+1}(x)]$. Let $\sum_{k=1}^{\infty} (g_k - g_{k+1}) = \varphi$. Then, $\| \varphi \|_p \leq \| F \|_p \leq 1$. But, $\sum_{k=1}^{n} (g_k - g_{k+1}) = g_1 - g_{n+1}$. So $\lim_{n \to \infty} g_{n+1} = g_1 - \varphi$.

Let $f = g_1 - \varphi$. We claim:

(a) $f \in \mathcal{L}_p$,
(b) $f = \lim_{n \to \infty} g_n$ a.e.,
(c) $\| f - g_n \| \to 0$ as $n \to \infty$.

Part (a) is trivial because $g_1 \in \mathcal{L}_p$, and $\varphi \in \mathcal{L}_p$, part (b) is the definition of f, and part (c) is shown as follows: from the above we have $| g_n - g_1 | \leq F$ and $| f - g_1 | \leq F$. Therefore,

$$| f - g_n | \leq | f - g_1 | + | g_1 - g_n | \leq 2F.$$

Thus, $| f - g_n |^p \leq 2^p F^p$, $n = 1, 2, \ldots$. But, $F \in \mathcal{L}_p$ and $| f - g_n |^p \to 0$ a.e. So, by Theorem 9.3.8,

$$\lim_{n \to \infty} \int | f - g_n |^p = \int \lim_{n \to \infty} | f - g_n |^p = \int 0 = 0.$$

So, $\| f - g_n \|_p \to 0$ as $n \to \infty$. ∎

PROBLEMS

1. Write out the proof of Theorem 9.4.2.

2. Complete the proof of Theorem 9.4.3.

3. Notation 9.4.1 is not valid unless we assume f is measurable. Give an example of a function which is not measurable but for

which $\int_E f^2$ exists and is finite. Such a function is not an element of L_2 nor is its equivalence class an element of $\mathcal{L}_2$.

4. Recall that convergence in the norm of $C[a, b]$, $C^{(1)}[a, b]$, and $B[a, b]$ implies pointwise convergence but that the converse is not true. (In fact, many of the problems in Chapters III–VI deal with the question of whether norm convergence is present when we obviously have pointwise convergence.) In $\mathcal{L}_p$ the situation is somewhat different as the following example shows: write each positive integer n as $n = 2^s + k$ (s and k integers, $0 \le k < 2^s$) and let

$$f_n(x) = \begin{cases} 1 & \text{if } k2^{-s} \le x < (k+1)2^{-s}, \\ 0 & \text{elsewhere.} \end{cases}$$

Show that $\| f_n - 0 \|_1 \to 0$ and $\| f_n - 0 \|_2 \to 0$, that is, $\int_E | f_n | \to 0$ and $\int_E | f_n |^2 \to 0$, where $E = [0, 1)$, but that $\{ f_n(x) \}$ does not converge to 0 for any $x \in E$.

5. Let $f \in \mathcal{L}_2([-\pi, \pi])$. Show that the integrals

$$\int_{-\pi}^{\pi} f(x) \cos nx \, dx \qquad \text{and} \qquad \int_{-\pi}^{\pi} f(x) \sin nx \, dx$$

exist for $n = 1, 2, \ldots$. (*Hint*:   use Hölder's inequality.)

6. Let

$$f_n(x) = \sum_{k=1}^{n} (a_k \cos kx + b_k \sin kx),$$

where $\sum_{k=1}^{\infty} | a_k |$ and $\sum_{k=1}^{\infty} | b_k |$ are finite. Show that there exists a function $f \in \mathcal{L}_2$ such that $\| f - f_n \|_2 \to 0$. Then show that

$$a_k = \frac{1}{\pi} \int_{-\pi}^{\pi} f(x) \cos kx \, dx \qquad \text{and} \qquad b_k = \frac{1}{\pi} \int_{-\pi}^{\pi} f(x) \sin kx \, dx.$$

(*Hint*:   evaluation of integrals like $\int_{-\pi}^{\pi} \sin nx \cos mx \, dx$ is needed.)

7. "Fourier series" is usually presented in differential equations courses from the point of view of Riemann integrals. Using

Problems 5 and 6, reformulate the major results from any standard text in differential equations in terms of functions in $\mathcal{L}_2$.

8. Using the definition of l_∞ as a guide and recalling that Lebesgue
integration does not distinguish between functions which are
equal a.e., formulate the definition of the space $\mathcal{L}_\infty$, including its
norm. Prove $\mathcal{L}_\infty$ is complete.

APPENDIX

Graphs on the following pages illustrate various types of norm convergence or lack of it.

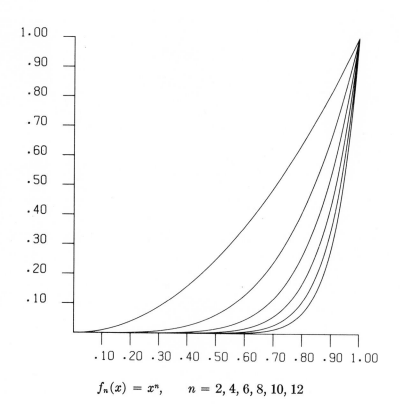

$$f_n(x) = x^n, \qquad n = 2, 4, 6, 8, 10, 12$$

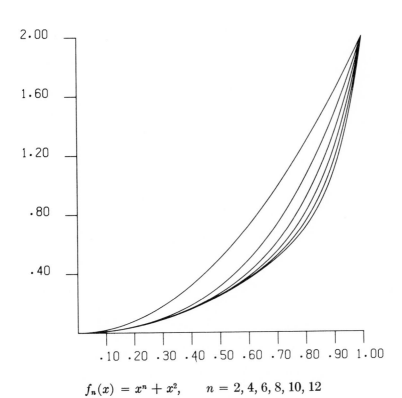

$$f_n(x) = x^n + x^2, \qquad n = 2, 4, 6, 8, 10, 12$$

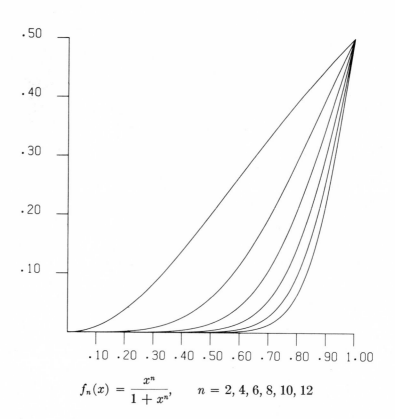

$$f_n(x) = \frac{x^n}{1 + x^n}, \qquad n = 2, 4, 6, 8, 10, 12$$

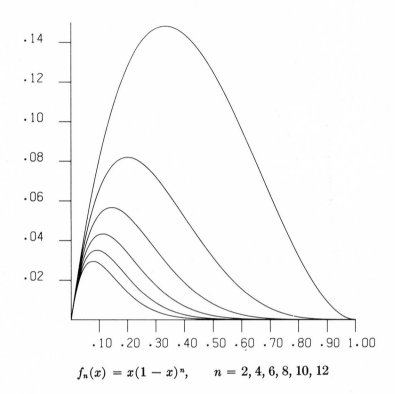

$$f_n(x) = x(1-x)^n, \qquad n = 2, 4, 6, 8, 10, 12$$

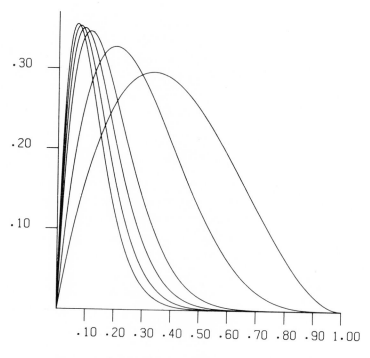

$$f_n(x) = nx(1-x)^n, \qquad n = 2, 4, 8, 10, 12, 14$$

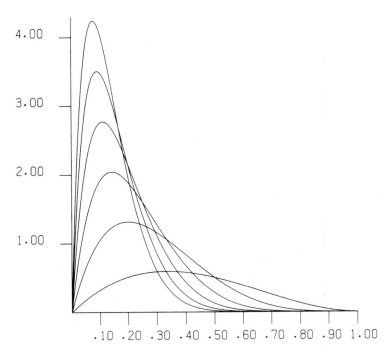

$$f_n(x) = n^2 x (1-x)^n, \qquad n = 2, 4, 6, 8, 10, 12$$

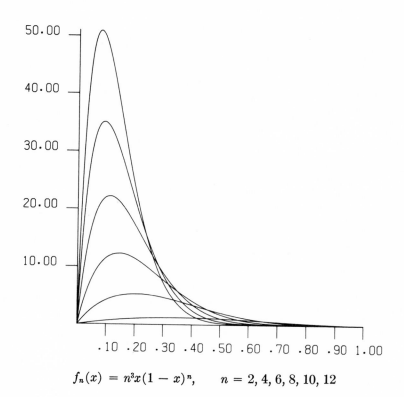

$$f_n(x) = n^3x(1 - x)^n, \qquad n = 2, 4, 6, 8, 10, 12$$

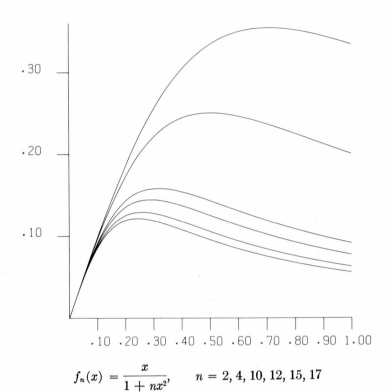

$$f_n(x) = \frac{x}{1 + nx^2}, \qquad n = 2, 4, 10, 12, 15, 17$$

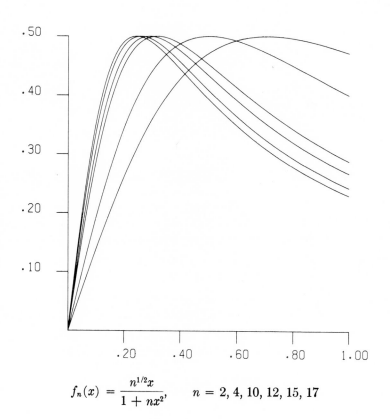

$$f_n(x) = \frac{n^{1/2}x}{1 + nx^2}, \qquad n = 2, 4, 10, 12, 15, 17$$

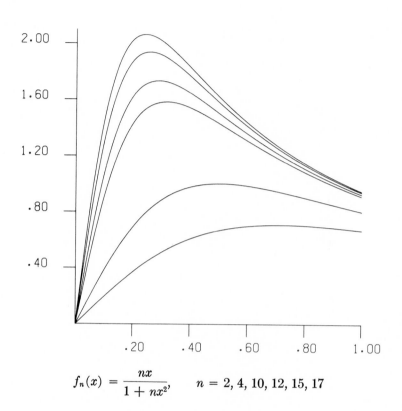

$$f_n(x) = \frac{nx}{1 + nx^2}, \qquad n = 2, 4, 10, 12, 15, 17$$

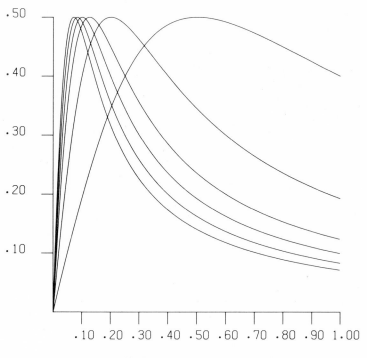

$$f_n(x) = \frac{nx}{1 + n^2x^2}, \qquad n = 2, 5, 8, 10, 12, 14$$

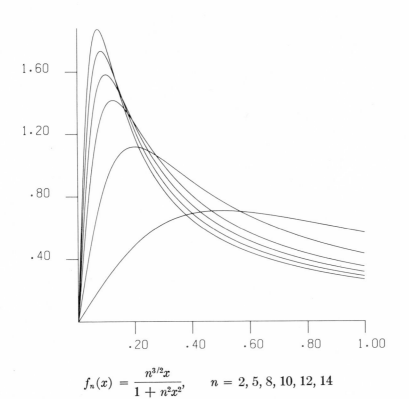

$$f_n(x) = \frac{n^{3/2}x}{1 + n^2x^2}, \qquad n = 2, 5, 8, 10, 12, 14$$

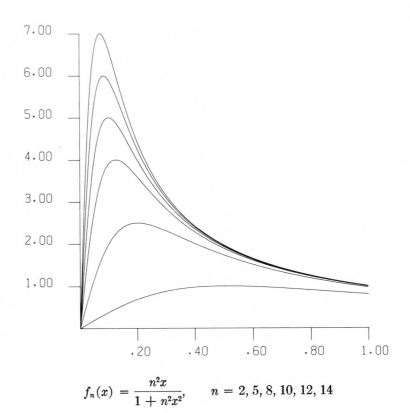

$$f_n(x) = \frac{n^2 x}{1 + n^2 x^2}, \qquad n = 2, 5, 8, 10, 12, 14$$

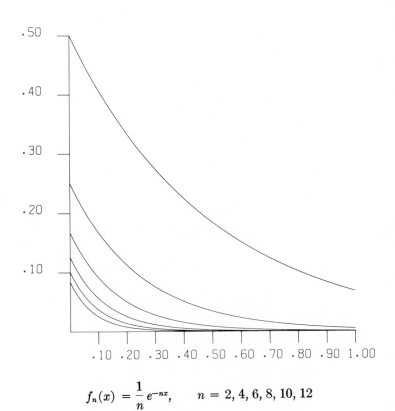

$$f_n(x) = \frac{1}{n} e^{-nx}, \qquad n = 2, 4, 6, 8, 10, 12$$

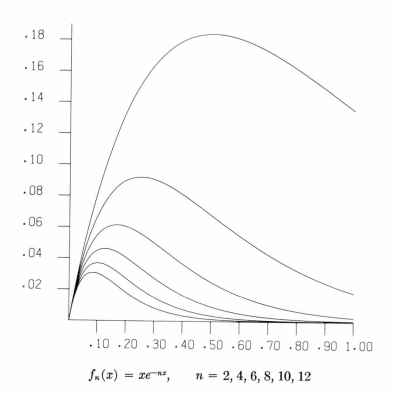

$$f_n(x) = xe^{-nx}, \qquad n = 2, 4, 6, 8, 10, 12$$

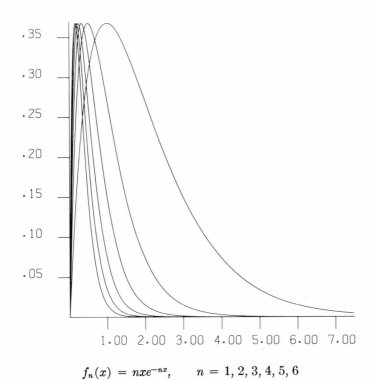

$$f_n(x) = nxe^{-nx}, \qquad n = 1, 2, 3, 4, 5, 6$$

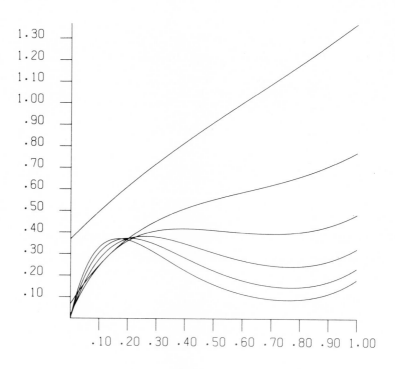

$$f_n(x) = nxe^{-nx} + \frac{e^{n(x-1)}}{n}, \qquad n = 1, 2, 3, 4, 5, 6$$

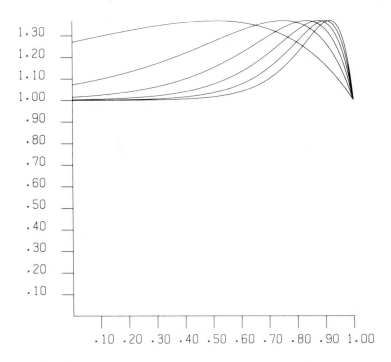

$$f_n(x) = ne^{n(x-1)}(1-x) + 1, \qquad n = 2, 4, 6, 8, 10, 12$$

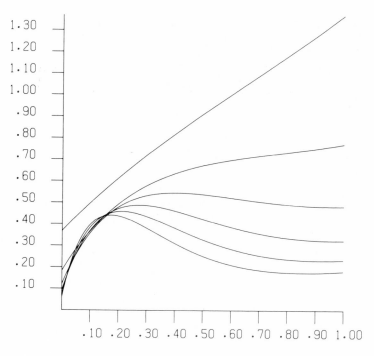

$$f_n(x) = nxe^{-nx} + \frac{e^{x-1}}{n}, \qquad n = 1, 2, 3, 4, 5, 6$$

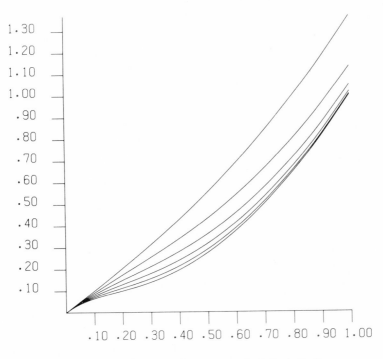

$$f_n(x) = x(e^{-nx} + x), \qquad n = 1, 2, 3, 4, 5, 6$$

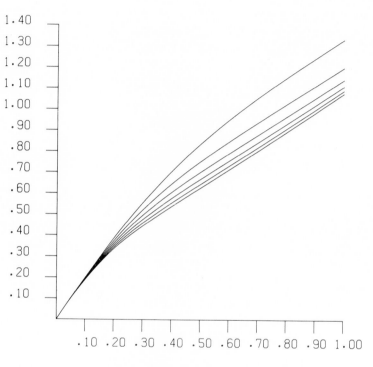

$$f_n(x) \ = \ \frac{2x \ + \ nx^3}{1 \ + \ nx^2}, \qquad n \ = \ 2, 4, 6, 8, 10, 12$$

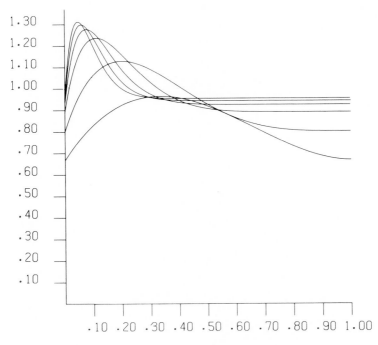

$$f_n(x) = n\left\{x(1-x)^n + \frac{1}{n+1}\right\}, \quad n = 2, 4, 8, 12, 16, 20$$

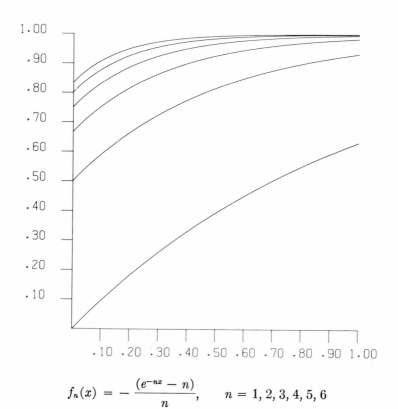

$$f_n(x) = -\frac{(e^{-nx} - n)}{n}, \qquad n = 1, 2, 3, 4, 5, 6$$

SUGGESTIONS FOR FURTHER READING

The following list is by no means an exhaustive bibliography. It contains books referred to in the text and a few books for reference or parallel reading.

APOSTLE, T. M., "Mathematical Analysis." Addison-Wesley, Reading, Massachusetts, 1957.

DIEUDONNÉ, J., "Foundations of Modern Analysis," enlarged and corrected printing. Academic Press, New York, 1960.

DUNFORD, N., and SCHWARTZ, J. T., "Linear Operators, Part I." Wiley (Interscience), New York, 1958.

HALMOS, P., "Measure Theory." Van Nostrand-Reinhold, Princeton, New Jersey, 1950.

HALMOS, P., "Naive Set Theory." Van Nostrand-Reinhold, Princeton, New Jersey, 1960.

HILLE, E., "Analytic Function Theory," Vol. 1. Ginn (Blaisdell), Boston, Massachusetts, 1959.

KELLEY, J., "General Topology." Van Nostrand-Reinhold, Princeton, New Jersey, 1955.

LANDAU, E., "Foundations of Analysis," 2nd ed. Chelsea, Bronx, New York, 1960.

SPRECHER, D. A., "Elements of Real Analysis." Academic Press, New York, 1970.

TAYLOR, A. E., "Advanced Calculus." Ginn (Blaisdell), Boston, Massachusetts, 1955.

WHITTAKER, E. T., and WATSON, G. N., "A Course of Modern Analysis," American ed. Macmillan, New York, 1943.

SOLUTIONS TO SELECTED PROBLEMS

CHAPTER I

Section 1.1

7. $\left(h \circ (f + g)\right)(t) = |\,t^2 - 1\,|$; domain $(-\infty, \infty)$; range $[0, \infty)$.

9. $(f \circ g)(t) = \sin(2\cos 2t)$; domain $(-\infty, \infty)$; range $[-1, 1]$.

11. $f(t) = \dfrac{1}{t}$.

13. $f(t) = \begin{cases} 0 & \text{if } t = 0, \\ 1 & \text{if } 0 < t \leq \dfrac{\pi}{2}. \end{cases}$

15. $f(t) = 0$ for $1 \leq t < \infty$.

17. $(-1, 1)$.

19. $(-\infty, \frac{3}{4})$.

21. $C[-1, 1]$.

23. None.

25. None.   **29.** $C^{(1)}[0, \pi]$.

27. $C[-1, 1)$.

Section 1.2

1. (a) $[-1, 1]$;   (b) inside or on the unit circle.

3. Yes; on a square joining the *points* $(1, 0)$, $(0, 1)$, $(-1, 0)$, and $(0, -1)$.

5. 24; yes.

11. No; $\sup_{0 \le t \le \frac{1}{2}\pi} |f_n(t) - f(t)| = 1 \nrightarrow 0$.

17. Yes.

Section 1.3

5. In the first octant, the plane determined by $(1, 0, 0)$, $(0, 1, 0)$, $(0, 0, 1)$; similarly for the other octants.

7. Let

$$H = 1 + \frac{1}{2} + \frac{1}{3} + \cdots \frac{1}{k}.$$

Then

$$\| x_1 \| = 2H,$$

$$\| x_2 \| = H + 1 + \frac{1}{2^2} + \cdots + \frac{1}{k^2},$$

$$\| x_n \| = H + 1 + \frac{1}{2^n} + \cdots + \frac{1}{k^n}.$$

9. $\| x_1 \| = \left(2^2 + 1^2 + \left(\frac{2}{3}\right)^2 + \cdots + \left(\frac{2}{k}\right)^2 \right)^{1/2}.$

$\| x_2 \| = \left(2^2 + \left(\frac{3}{4}\right)^2 + \left(\frac{4}{9}\right)^2 + \cdots + \left(\frac{k+1}{k^2}\right)^2 \right)^{1/2}.$

$$\| x_n \| = \left(2^2 + \left(\frac{2^{n-1} + 1}{2^n} \right)^2 + \cdots + \left(\frac{k^{n-1} + 1}{k^n} \right)^2 \right)^{1/2}.$$

11. $\| x_1 \| = \| x_2 \| = \cdots = \| x_n \| = 2.$

CHAPTER II

Section 2.1

3. $\lim a_n = 0.$

5. $\limsup a_n = \ln 2, \liminf a_n = -\ln 2$; both are limit points.

7. $\limsup a_n = 1, \liminf a_n = -2$; 1, -2, and 0 are limit points.

9. $\lim a_n = 0.$

11. $\limsup a_n = 4, \liminf a_n = -\frac{3}{2}.$

13. $\lim a_n = 0.$

15. Any sequence $\{a_n\}$ for which $\lim a_n$ exists and equals $-\infty$ will do.

17. $a_n = 2n - 1; a_n = 6 - \frac{65}{6}n + 7n^2 - \frac{7}{6}n^3.$

Section 2.2

1. 2, 3, 4, 9, 13 are in (c_0); those and 6 are in (c); all are in l_∞.

7. $x = \left(1, \frac{1}{2}, \frac{1}{3}, \ldots, \frac{1}{k}, \ldots \right).$

9. $e_i = (\delta_{i1}, \delta_{i2}, \ldots)$, where $\delta_{ij} = 1$ if $i = j$; $\delta_{ij} = 0$ if $i \neq j$. No; also need $(1, 1, \ldots).$

Section 2.3

9. $\frac{1}{3}(\frac{1}{7})^3.$ **11.** 9 terms.

13. $\frac{8}{7}(\frac{1}{3})^7$.

15. $|R_n| \leq \dfrac{|a_{n+1}|^2}{|a_{n+1}| - |a_{n+2}|}$.

Section 2.4

5. Hint: if $\lim_{n\to\infty} a_{n+1}/a_n = L$, let $0 < s < L < t$. Then for some N,

$$s a_N < a_{N+1} < t a_N,$$
$$s^2 a_N < s a_{N+1} < a_{N+2} < t a_{N+1} < t^2 a_N,$$

or

$$s^p a_N < a_{N+p} < t^p a_N.$$

Let $N + p = n > N$; recall that $\liminf \sqrt[n]{a_n}$ and $\limsup \sqrt[n]{a_n}$ always exist and that $\liminf \leq \limsup$.

9. Convergent.

11. Absolutely convergent.

13. Convergent.

15. Convergent.

17. Divergent.

19. Convergent.

21. Conditionally convergent.

23. Convergent.

25. Convergent.

27. Conditionally convergent.

29. Convergent.

31. Convergent.

33. Divergent.

Section 2.5

1. $0 \leq x < 4$.

3. $-e < x < e$.

5. $-7 \leq x < 1$.

7. $0 < x$.

9. $-\infty < x < \ln 2$.

11. $0 < x < \infty$, also $-\infty < x < -2$.

13. (a) $1 + \dfrac{1}{x}$;   (b) $\dfrac{1}{1 - e^x}$.

Section 2.6

4. See Apostle, *Mathematical Analysis*, pp. 376, 377.

5. $\frac{3}{2}S$ has been rearranged.

9. Use Hölder's inequality.

CHAPTER III

Section 3.2

1. $f(x) \equiv 0; f \in C[0, 1]; \lim_{n \to \infty} \|f_n - f\| = \frac{1}{2}$.

3. $f(x) \equiv 0; f \in C[0, 1]; \lim_{n \to \infty} \|f_n - f\| = 1/e$.

5. $f(x) \equiv 0; f \in C[0, 1]; \lim_{n \to \infty} \|f_n - f\| = 0$.

7. $f(x) \equiv 0; f \in C^{(1)}[0, 1]; \lim_{n \to \infty} \|f_n - f\| = 1$.

9. $f(x) \equiv 0; f \in C[0, \infty); \lim_{n \to \infty} \|f_n - f\| = 1/e$.

13. Yes, in both cases.

17. 3 is, 4 is not.

CHAPTER IV

Section 4.1

7. (a) Define $f(2) = 12$; (d) $\lim_{x \to 0+} \neq \lim_{x \to 0-}$.

11. (a) No; (c) no; (e) yes.

13. (b) Let $\delta < \epsilon/2M$.

15. $x = \frac{1}{2}$.

Section 4.2

1. $c_n = 3^{-1/n}$; $\lim_{n \to \infty} c_n = 1$, but $f(1) \neq \frac{1}{4}$; f is not continuous.

3. $\lim_{x \to 0+} f(x) = 1$, $\lim_{x \to 0-} f(x) = -1$, $f(0) = 0$.

11. No.

13. $nxe^{-nx} \to 0$ but not uniformly; $e^{x-1}/n \to 0$ uniformly, so convergence of the sum is not uniform.

Section 4.3

3. (a) Any closed interval;   (b) any finite set;   (c) $\{1/n\}$.

5. In E^1, the rational numbers, the irrational numbers, and all the real numbers.

9. (a) Closed [note that $(0, 0)$ is an isolated point];   (b) open;   (c) neither.

Section 4.4

3. Let $y = mx$. Then $\lim_{x \to 0} f(x, mx)$ depends on m.

5. 0 (see Example 4.4.4).

7. 0 (note that $\delta^2 \ln 2\delta^2 \to 0$).

CHAPTER V

Section 5.2

3. $f(x) \equiv 1$; no; yes.

5. $f(x) \equiv 0$; no; yes.

7. $f(x) \equiv x^2$; no; yes.

Section 5.4

1. $|R_n| \le e^\xi x^n/n! \to 0$ as $n \to \infty$ for every x.

5. Consider $x^n \sin(1/x)$ for $n = 1, 2, \ldots$ and see what values of n will give the conditions required.

9. $x^4 = 1 + 4(x - 1) + 6(x - 1)^2 + 4(x - 1)^3 + (x - 1)^4$.

CHAPTER VI

Section 6.1

7. (a) Yes; (b) no; (c) yes; (d) no.

9. $6\frac{1}{2}$.

11. $(\sin x)/(x + 1)$ is increasing for $0 \le x \le 1$. Thus,

$$\underline{s_n} = 0 + \frac{\sin(1/n)}{1 + n} + \frac{\sin(2/n)}{2 + n} + \cdots + \frac{\sin[1 - (1/n)]}{2n - 1}$$

$$\overline{S_n} = \underline{s_n} + \frac{\sin 1}{2n}.$$

13. $\underline{s_n} = \frac{1}{n} + (e^{1/n} - 1) + \frac{e^{2/n} - 1}{2} + \cdots + \frac{e^{1-(1/n)} - 1}{n - 1},$

$$\overline{S_n} = \underline{s_n} - \frac{1}{n} + \frac{e - 1}{n}.$$

15. Lower sum for $\displaystyle\int_0^1 \frac{dx}{1 + x}$.

Section 6.2

1. $2xe^{-x^4} - e^{-x^2}$.

3. $4x^3 \sin x^2 - 2x \sin |x|$.

7. $\ln 2$.

Section 6.3

5. $p < 1$.   **13.** -1.

6. $p > 1$.   **19.** (a) $2 \int_0^\infty e^{-y^2} y^{2x-1} \, dy$.

9. Convergent.   **23.** $\dfrac{a}{s^2 + a^2}$.

11. Convergent.   **25.** $\dfrac{1}{s - a}$.

Section 6.4

1. (a) $\dfrac{n^2}{(n+1)(n+2)}$; (b) $f(x) \equiv 0$; (c) no; (d) no.

3. (a) $-e^{-n} - \dfrac{e^{-n}}{n} + \dfrac{1}{n}$; (b) $f(x) \equiv 0$; (c) no; (d) yes.

5. (a) $\dfrac{n^2}{(n+1)(n+2)} + \dfrac{1}{2}$; (b) $f(x) = x$; (c) no; (d) no.

7. (a) Yes; (b) $y = x^n$ will do.

CHAPTER VII

Section 7.1

3. Show that the infinite sum converges uniformly and then integrate.

5. Show that the infinite sum and the infinite sum of the derivatives both converge uniformly and then differentiate.

7. $f(x) = x$; yes; no.

9. $f(x) \equiv 1$; no; no.

Section 7.2

5. $x + \dfrac{x^3}{3} + \dfrac{2x^5}{15} + \dfrac{17x^7}{315} + \cdots; |x| < \dfrac{\pi}{2}.$

7. $x + \dfrac{x^3}{6} + \dfrac{1}{2} \cdot \dfrac{3}{4} \cdot \dfrac{x^5}{5} + \dfrac{1}{2} \cdot \dfrac{3}{4} \cdot \dfrac{5}{6} \cdot \dfrac{x^7}{7} + \cdots; |x| < 1.$

9. $-\dfrac{x^2}{2} - \dfrac{x^4}{12} - \dfrac{x^6}{45} - \dfrac{17x^8}{2520} - \cdots; |x| < \dfrac{\pi}{2}.$

11. $\dfrac{1}{3!} - \dfrac{x^2}{5!} + \dfrac{x^4}{7!} - \cdots$; series converges for all x, but it converges

to the function given only if $x \neq 0$.

CHAPTER VIII

Section 8.1

1. (a) $[-3, -2] \cup [2, 3]$; (b) $(-1, 1)$.

3. $\{-1\} \cup [2, 3)$. **7.** Measure is zero. **9.** $\frac{1}{2}$.

Section 8.2

1. Should know that an open set is measurable according to Definition 8.2.3.

11. One.   **13.** No.

CHAPTER IX

Section 9.1

5. Let U be the nonmeasurable set in $[0, \pi]$ described in Section 8.3. Let

$$f(x) = \begin{cases} 1 & \text{if } x \in U, \\ 0 & \text{if } x \in (\pi, 2\pi], \\ -1 & \text{if } x \in [0, \pi] - U. \end{cases}$$

Then the domain of f is $[0, 2\pi]$.

$\{x \mid f(x) = 0\}$ is measurable
$\{x \mid f(x) < 0\}$ is not measurable
$\{x \mid f(x) > 0\}$ is not measurable.

7. $f \circ g$ is not necessarily measurable.

Section 9.4

3. The function described for Problem 5, p. 153 will do.

Subject Index